THE

SOLAR

PHOENIX

HOW AMERICA CAN RISE FROM THE ASHES OF SOLYNDRA TO WORLD LEADERSHIP IN SOLAR 2.0

BRAD MATTSON

The Solar Phoenix

© 2014 Brad Mattson

ISBN: 978-1-61170-177-7

Printed in the USA and UK on acid-free paper.

 **Robertson Publishing**™
www.RobertsonPublishing.com

To purchase additional copies of this book go to:
amazon.com
barnesandnoble.com

For Vicky

Maybe it's trite to dedicate a book to your wife,
but I'm just going to have to be trite.

I had no idea how much work this book was going to be,
but she did. She gave up a year's worth of weekends and a va-
cation in Hawaii (while I did nothing but write)… and we
never even set foot on the beach.

In the end, when I wasn't sure if I could finish, she stepped in
and became researcher, editor, publisher, cheerleader,
and project manager.

It is no lie when I say it could not have been done without
my partner, my best friend…my boss.

Table of Contents

~ ~ ~ ~ ~ ~ ~ ~ ~ ~ ~ ~

Acknowledgements

I had no clue what I was getting myself into when my longtime friend and colleague, Mary Curtis, suggested that I write this book. Considering I had little time and no writing expertise, *The Solar Phoenix* has been a group effort from the very beginning. The following people helped me refine the arguments and put them into words. I am so grateful for their contributions.

Markus Beck	Willie Brent	Mary Pacifico Curtis
Sherrill Dale	Mike Danaher	Craig Elliot
Les Fritzemeier	Melody Haller	Marianne L. Hamilton
Craig Leidholm	Vicky Mattson	Cait Murphy
Alicia Robertson	Brendan Ruiz	Fred Tabrizi
Robert Wendt	Eric Wesoff	David Williams

A number of researchers and analysts put together the data and allowed me to use it. Special thanks to: Bloomberg, Bright Side Network, DBL Investors, Deutsche Bank, EIA, European PV Industry Association (EPIA), GreenTech Media, IC Insights, IEEE, Keiser Analytics, Navigant Consulting, NREL, PV Tech, Rocky Mountain Institute, Solar Energy Industries Association (SEIA), SolarBuzz, The Smart Cube, Tigercomm, US Department of Energy, and Vote Solar.

I am bound to have forgotten someone who belongs on the list. I'll apologize in advance and yes, let's have lunch (my treat). Any mistakes or omissions are entirely my fault.

Preface

If you agree with me on 9 out of 12 issues, vote for me. If you agree with me on 12 out of 12 issues, see a psychiatrist. (Ed Koch, former mayor of New York)

When I read books, I often skip the preface. But I am glad that you are reading this because there are some things you should know.

The fact that I undertook the writing of this book while working as CEO of a solar startup probably proves my lunacy (to those not already convinced of it). I took this on because I am passionate about making solar energy succeed, and in a big way. This can and should happen. But it hasn't yet.

I believe that we are in the most exciting time in the history of solar; the industry is truly at a tipping point. The costs have come down so far that solar is now economical in most places around the world. At the same time, it is misunderstood by much of the public and out of favor with some investors due to the turmoil caused by recent industry consolidation. Even though solar has grown 50% a year for almost a decade, 80% of the manufacturers have gone out of business or been acquired over the same period. Closer to home, I fear that the US could miss its chance to benefit from this once-in-a-lifetime transformation to a new energy economy.

Why I Care

Like many of my neighbors in Silicon Valley, I spent the 1980s and 1990s chasing high-tech dreams. I was one of the lucky ones: Two of the startups I founded went public. In the wake of that success, I began to ask myself: "Has technology really made the world better?" To my surprise, I found I was not sure. It occurred to me, for example, that technology played a role in widening the global gap between rich and poor. What does the hoopla over the next generation of tablets mean to those who have no access to clean water, basic sanitation, or lighting?

Around 2004, I decided to use my business and technology skills to help bridge these gaps. Through my work with the Global Social Benefit Incubator (GSBI) at Santa Clara University, I mentored entrepreneurs in developing countries that were using science, technology, and unique business models to address the challenges of poverty. Many of these innovations were fascinating. Kiva, the micro-lending site, is one graduate. HuskPower, which provides low-cost, bio-based electricity to Indian villages, is another. But it was solar power that inspired me most. I saw how essential affordable energy is to economic development, and that solar could help. Here, I came to believe, is a technology that can improve lives all over the world.

In 2008, with the idea of helping solar to scale up and get cheaper, I joined a venture capital firm, VantagePoint Capital Partners, which made some significant solar investments. Two years later, I became CEO at Solexant (now Siva Power), which is developing what we believe is the next generation of solar panel technology.

In short, I have spent a number of years studying the industry, in terms of investment, operations, technology, and business models. As a businessman and an engineer, I am only interested in things that work and that have sufficient returns on investment to make them scalable and sustainable. I know the reality. I have gone through the pain of laying off valued employees, and the frustration of trying to raise money from investors who are disappointed in the sector.

I have learned a lot — and I still have a lot to learn. But I am optimistic. Solar is financially viable, and the world needs clean and plentiful sources of energy. It is time.

Yes, I have an agenda. *The Solar Phoenix* aims to build support for solar, by laying out the facts. By reviewing the data and clarifying some myths, I believe the case makes itself. *The Solar Phoenix* is meant for a general audience, not a specialized one, and while the opportunities in solar will bring global benefits, I discuss the US in greater detail. I think the US should exercise its strengths in technology and innovation to become the world leader in solar.

At the beginning of each chapter I summarize the most important points, so you can still get the main takeaways without diving into the details. Feel free to skip around.

A final note. The idea of this book is to start a conversation. So please start talking — with your friends, colleagues, family — and me

- on Twitter @bradmattson
- or through the book website: thesolarphoenix.com

You may not agree with me 12 out of 12 times, as Mayor Koch put it. But when you finish, I hope you will share my conviction that solar is a technology that can benefit humanity.

There's nothing crazy about that.

THE

SOLAR

PHOENIX

Chapter 1

Why Energy Matters

Why does it matter how we warm our homes or power our cars? Because how energy is generated affects the health of every human on earth, and the economies of every nation. Energy is a regular flashpoint in global conflict, and it also plays a crucial role in the well-being of the planet, particularly in the effort to combat climate change. So energy matters — everywhere and all the time. But while energy policy is widely debated, too often these discussions generate more heat than light. In this chapter I discuss the current energy situation, and make the case that the time for renewable energy has come.

In an off-grid Eastern African village, a mother (who once paid 30% of her income for kerosene and candles to light her hut) watched her children do their evening homework by the light of a solar lantern that costs nothing to recharge each day. In rural India, a shop owner increased his income more than 20% by staying open in the evenings, thanks to lighting from a solar-powered mini-grid in his remote village. On the Caribbean coast of Nicaragua, a doctor was finally able to prevent childhood diseases after the installation of solar panels to power a small refrigerator for vaccines.

Meanwhile, here in Silicon Valley, where people camp out overnight to buy the latest iPhone, most people don't think much about where their energy comes from. In fact, most of us in the Western world may grumble when paying our electrical bills, but we take for granted that electricity will be readily available to power everything we need and want to live, work, and play. Why should we care about where that electricity comes from? There are, in fact, several good reasons to care.

Health Consequences

One reason is the indisputable and serious health impact of burning fossil fuels for energy. David Pimentel, professor of ecology at Cornell,[1] estimates that air pollution from smoke and various chemical emissions kills 3 million people each year. Coal is the dirtiest offender, and carries the most serious health consequences.[2] So it is unfortunate, and little short of tragic, that coal is also the dominant source of energy, accounting for about 25% of global production.

The problem is particularly acute in China (see Fig. 1.1). Take a look at *China Air Quality Index*, a popular iPhone app that reports on air quality for the major cities in China, essentially sounding an alert for "pollution days." Kids stay home from school, and some people stay home from work. Consider the potential for civic unrest when citizens' lives are threatened by the air that they breathe. That is why China is putting serious money behind renewable energy. The stability of Chinese society may require it.

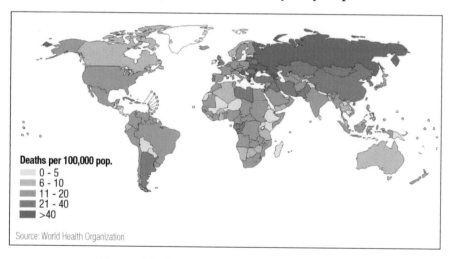

Deaths per 100,000 pop.
- 0 - 5
- 6 - 10
- 11 - 20
- 21 - 40
- >40

Source: World Health Organization

Figure 1.1 - Deaths from urban air pollution

Energy Independence

The political and economic importance of energy is undeniable. Throughout modern history, countries—including, of course, the US—have fought wars over energy. Every nation wants a secure source of energy, which is crucial to economic and military capabilities, and they are willing to pay dearly for it, in lives and dollars. If more countries could move toward energy independence—meaning that they control their own sources of energy—there could be significant and positive changes in international relations. This could be called the "democratization of energy," or energy available to everyone.

On the flip side, lack of energy independence can have serious financial implications. There is the cost of war, if it comes to that, or the cost of maintaining a military presence in oil-producing regions, or simply the cost of using hard currency to buy foreign energy supplies. In the US, our balance of trade and our budget deficit are affected by our lack of energy independence. In the end, we pay for all of this with our taxes. So there are both deep economic and human costs to America's lack of energy independence.

Lack of energy independence hurts any country that doesn't have an energy surplus, which happens to be most of them. And it is not even clear in the long run if it helps those countries that are exporting energy. Just look at Nigeria, Venezuela, the Middle East, or Russia. Exporting their cheap fossil fuels hasn't led to a competitive domestic economy yet. In fact, cheap natural resources may be a curse, not a blessing for a developing nation. In a quote from his article, "Natural Resource Abundance and Economic Growth," Jeffrey Sachs put it this way: "One of the surprising features of modern economic growth is that economies with abundant natural resources have tended to grow less rapidly than natural-resource-scarce economies".[3] Yes, the democratization of energy, energy independence, in my humble opinion, could be a good thing for everyone, everywhere.

Climate Change

Then there is climate change. There is a strong correlation between CO_2 emissions and the surface temperature of the earth (see Fig. 1.2). While most of the CO_2 in the atmosphere is naturally occurring, man-made sources of CO_2 may be upsetting the earth's delicate balance. This could mean bizarre weather patterns,[4] rising sea levels, shrinking glaciers, and other ecosystem damage. While climate change is not fully understood, it is clear that non-fossil fuel forms of energy are needed to contain and eventually reduce greenhouse-gas emissions.

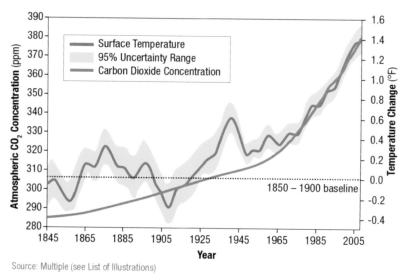

Source: Multiple (see List of Illustrations)

Figure 1.2 - Atmospheric CO_2 vs. surface temperature

Diminishing Supplies

Finally, there is the simple fact that while there are still large reserves of fossil fuels (enough coal for more than 100 years, for example, according to the US Energy Information Administration), these are not inexhaustible. Moreover, some kinds of fossil fuels are getting tougher to find, and therefore are more expensive. Even the major energy companies, for example, are seeing profits hit due to higher capital costs related to finding and extracting oil; Shell reported in early 2014 that its cost to pump a barrel of oil had risen 56% in the previous three years.[5]

A former Saudi oil minister, Ahmed Zaki Yamani, noted in 2000, "The Stone Age came to an end not for a lack of stones, and the Oil Age will end, but not for a lack of oil." It makes sense to prepare for a future in which reserves diminish and extraction costs go up, raising the price of conventional energy. This is not theoretical; oil prices have risen from $38 a barrel in 2000 to almost $112 at the end of 2013.[6] Over the same period, the costs of renewables have gone way down, in the case of solar panels from $3.60/watt down to $0.60/watt. This is almost a 3X **increase** for oil versus a **6X decrease** for solar. At some point, the cost curves cross over and renewables will be at a cost advantage—even without factoring in the environmental or health costs of carbon-based fuels.

The end result is clear. The world cannot continue to rely on fossil fuels, and is going to need renewable sources to meet its energy needs.

Chapter 2

The Case for Renewable Energy

Energy is essential for economic growth: If we want people in places like Africa, India and Nicaragua to have lives of greater opportunity, they will need more energy. According to the UN, if residents of China and India eventually consume as much energy per capita as Americans, they will need 4 times and 14 times as much power, respectively, as they currently consume.

Looking at Figure 2.1, it is clear that energy and GDP go hand in hand. As the world's economies all want to catch up with the leaders (top-right on the chart), the increase in energy consumption will be enormous.

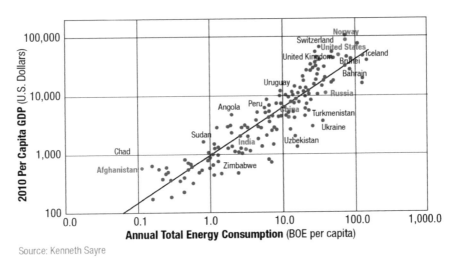

Source: Kenneth Sayre

Figure 2.1 – Per capita GDP vs. energy consumption

According to the UN, if residents of China and India (shown near the middle of the chart) eventually consume as much energy per capita as Americans, they will need 4 times and 14 times as much power, respectively, as they currently consume.[7]

The scale of the need is mind-boggling. The data in Figure 2.2 shows that by 2050 the gap between energy demand and the energy supply from conventional sources is equivalent to 40 billion barrels of oil. This translates into about 46,000 gigawatts (GW) of demand that would have to be met by renewable energy.[8] (A gigawatt can power about 750,000 American homes.) By comparison, in 2012 the world had added only about 200GW of renewable energy. That means we need to scale up by over 200 times to meet demand, adding more than 1,000GW per year every year for the next 40 years. We need to examine all viable energy sources to fill this huge gap.

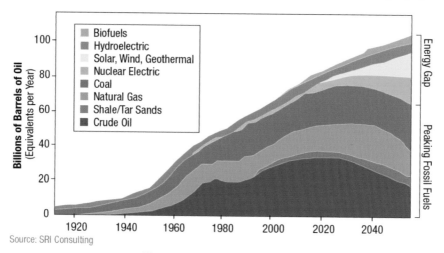

Figure 2.2 – The energy gap

Essentially, there are three broad options as shown in the energy ecosystem in Figure 2.3. **Conventional** energy sources refer to fossil fuels such as oil, coal, and gas that are dug out of the earth or sea; when they are used, they create greenhouse-gas emissions. Oil is used chiefly for transport; coal and gas for heating, lighting, and industry. **Renewables** refer to sources that create little or no emissions, and that are essentially endless: hydro, wind, solar, biofuels (biomass), and geothermal. **Nuclear** is nuclear.

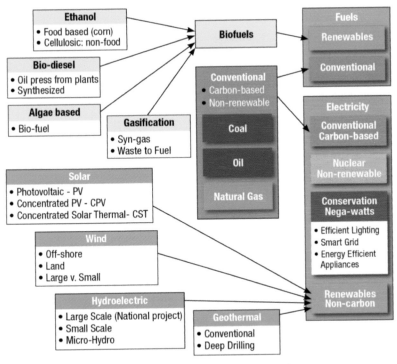

Figure 2.3 – The energy ecosystem

Right now, the great majority of the world's electricity is generated by conventional sources that are carbon-based and non-renewable. Unfortunately, as shown in Figure 2.4, coal, the dirtiest source, dominates, followed by natural gas, nuclear, and hydropower.

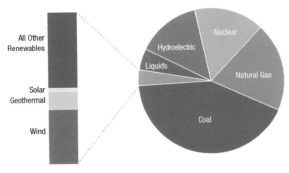

Source: EIA Data 2011 in International Energy Outlook 2010

Figure 2.4 – Global sources of electricity

That brings us all the way around the chart in Figure 2.4 to that little green sliver – non-hydro renewables. At first glance the chart makes renewables look, well, trivial. Solar, for example, barely registers.

But that is not the only way to consider things. Figure 2.4 is just a snapshot; if it were a moving image, the story would be very different. Specifically, in many places, renewables represent most new capacity. In January, 2013, for example, *all* new energy capacity added in the United States came from renewable sources.[9] Okay, that was a particularly good few weeks, but for the year as a whole, it was close to 30 percent, and more than half in some states. So although the green renewable sliver is small, the percentage of new capacity added is large. In other words, the renewable piece of the pie is getting bigger faster.

Given this context, how do renewables fit into the future of energy? Coal is abundant, but dirty and dangerous. Oil is getting more expensive, and because it is used almost entirely for transport, is not in the same market as renewables, which are used almost entirely to generate electricity. Natural gas is intriguing. This is the cleanest form of conventional energy, and supplies (particularly in the US) have been growing due to advanced extraction techniques, such as fracking.

Many environmentalists see the burgeoning of lower-cost natural gas in the US as a problem for renewables; I think they can work together. Natural gas could and should be viewed as a US national treasure—not by burning it, but by selling it. A likely long-term price for US natural gas (Fig. 2.5) is $5 to $6 per million BTU (mbtu); around the end of 2013, the spot price in Europe was around $10/mbtu and in Asia about $16/mbtu. That is an enormous opportunity. In what I call the "export scenario," the US sells natural gas for a premium abroad as it scales up the use of renewables to generate electricity at home.

Natural gas and renewables thus combine to create a clean and economically sustainable energy future. On this basis, perhaps greens and gas advocates should kiss and make up.

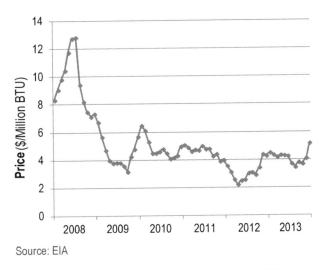

Source: EIA

Figure 2.5 – Natural gas prices in the US

In an oddly parallel situation, OPEC countries like Saudi Arabia are finally realizing it is economically foolish to use oil domestically to generate electricity. They can make so much more money exporting their oil that the opportunity costs become enormous. Saudi Arabia plans to spend $100 billion on solar electricity to minimize internal oil use and maximize exports.

It is, of course, essential to address the environmental side-effects of fracking, such as water-use and methane emissions. Still, natural gas could be an important source of cleaner energy during the transition to renewables. In the US, for example, natural gas has already displaced a significant amount of coal, which is a large part of the reason that greenhouse-gas emissions in 2012 were less than they were in 1994. Natural gas is, however, still a fossil fuel and subject to all the fossil fuel issues discussed above; it is not a long-term global solution for many reasons, including the fact that not everyone has natural gas reserves.

What about nuclear? Well, it's complicated. Nuclear is not considered a conventional source of energy. But even though it is essentially inexhaustible and emits no greenhouse gases, it is not considered renewable, either. Why not? Probably because current plants, which use fission, generate waste that takes centuries to break down. Nuclear fusion is elegant and clean. It's just that it doesn't exist. Fifty years ago, fusion was the technology that was "50 years away"; it may still be that far away now.

Nuclear is an important and needed source of energy (about 12% of global supply), but it is unlikely to supply the 46,000GW needed. In many countries, it is impossible to get permission to build new plants, which are also very expensive. In addition, the big issue of where and how to store nuclear waste has not been resolved. So although some countries are building more nuclear plants,[10] these aren't sufficient to bridge the energy gap.

The Case for Renewables

Coal is abundant, but has totally unacceptable health and environmental impacts; natural gas is much better, but produces carbon and will eventually run out; nuclear is just too unpopular. Meanwhile, the world needs more and more energy. The case for renewables is simple; in the long run it is the only option that works. And its big disadvantage — cost — is becoming less and less of an issue.

Using more renewables will improve human health and slow climate change. Also, because renewable energy sources are available worldwide and not controlled by a small number of nations, their use could minimize conflicts over energy. Renewables can help countries and individuals control their own energy destinies.

We need more renewables, and we need them now. But do we pursue an "all-of-the-above" energy strategy, or a more targeted one? The needs are so great, we have to pursue all of the above, but I believe solar has the sunniest future. Its main drawback to date has been the high initial cost of building the solar power plant; it is still the most expensive renewable. Reducing these costs is the key to solar's future—and it is happening (see Chapter 5). Finally, solar has one significant advantage—it works best with the existing electricity infrastructure, known as "the grid." That is the subject of the next chapter.

What Is Renewable Energy?

Renewable energy sources are those that are not diminished with use, or are replaceable within a short period of time. Here are the major forms:

Hydroelectric: Think dams. Hydro is by far the largest source of renewable energy, equivalent to 831 million TOE globally in 2012 (compared to 237 million for other renewables).[11] (TOE stands for tons of oil equivalent and is the amount of energy generated by burning one metric ton of oil.) Hydro is clean and can be cheap, but in many countries there is little new development because of environmental concerns about dams.

Wind: Think really big windmills. This is the second-largest global source of renewable energy (117 million TOE), and capacity is growing fast (18.1% in 2012). Wind is clean and can be economical, especially at large scale (1 megawatt or more; 1 megawatt can power about 300 American homes). In many places wind is now competitive with conventional power, or getting close.

Geothermal: Think hot, really deep holes. Water is pumped into the hole to create steam that is used to drive a turbine and generate electricity. Unlike wind and solar, geothermal provides a constant source of electricity, known as "baseload." One limitation is that geothermal is currently only economical in certain geological areas.

Solar: Think panels, either on rooftops or in massive arrays known as "solar farms." This book focuses on photovoltaic (PV), in which sunlight is converted into electricity by using solar panels. Solar energy plays a small part in the global energy system—just 21 TOE. But it is growing fast—58% worldwide in 2012—and 138% in the US.

Conservation: Think of energy not used. The unit of measure is the "negawatt." While negawatts cannot turn on the lights, what they can do is reduce future demand. In California, which is a leader in conservation, per capita energy demand is 40 percent below the US average. As awesome as negawatts are, they have their limits. Most new energy demand will come from emerging and growth countries. Because they don't use much electricity to begin with, conservation can't make as big a difference.

Chapter 3

The Grid vs. Distributed Generation

The grid refers to the infrastructure that generates and then delivers power from large centralized plants — mostly coal, gas, and nuclear — to users everywhere. The grid needs to be reliable, economical, stable, and capable of delivering electricity everywhere, all the time. Today the great majority of power is delivered through the grid, but there is a growing trend toward "distributed generation" (DG). This refers to power that is produced and used on-site or nearby, using sources that work on a smaller scale, such as wind, natural gas, solar, and biomass. The growth of DG will mean changes in the way utility companies — the rulers of the grid — need to operate.

The Grid

If electricity is the world's energy currency, the grid is its banking system. A typical grid (see Fig. 3.1) has three parts: generation (creating the power); transmission and distribution (sending it over the wires); and end users (turning on the lights in homes, offices, and industry).

Just as most of us don't think twice about what makes an ATM work, most of us take electricity for granted, only thinking about it if there is a power disruption. But let's take a few minutes to define what people want from a power system.

Cost is always important, and the cost of power varies significantly based on location. Electricity is sold in units of "kilowatt hours," (kWh) which is the amount of power it takes to light a 100-watt bulb for 10 hours. An average American household uses 720kWh a month.[12] In Kuwait consumers pay just a penny per kilowatt hour, while Solomon Islanders pay 86¢/kWh.[13] In the United States, 2012 costs ranged from a low of about 8.37 cents per/kWh in Louisiana to 37.3 cents in Hawaii.[14] These variations result from different costs of generation and transmission, as well as the degree of competition, level of subsidies, and attitudes of rate-setting commissions. Whatever the final cost, consumers tend to take electricity for granted and obviously do not want to see prices go up.

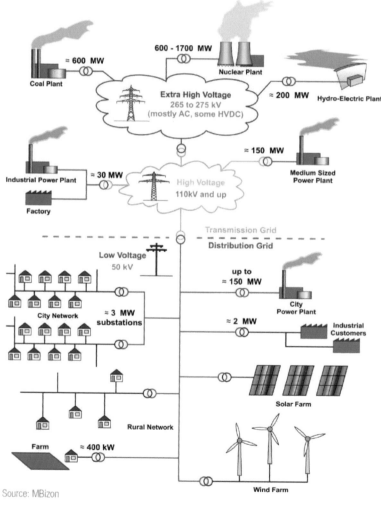

Source: MBizon

Figure 3.1 – Typical grid with renewable energy

Reliability is another essential; failure means power outages. It's important to recognize that the grid is not inherently stable. When someone consumes electricity, that power must be put back onto the grid at a generating plant to keep the system balanced. When consumption and generation changes do not balance—for example, if whole neighborhoods turn on their air conditioners at the same time—the grid can and does crash. A reliable system, then, requires power sources that can be turned on or off as needed, a require-ment, known as "dispatchability." When all those air conditioners go on, more power is needed—fast. Dispatchable generators, also known as "peak power" plants, respond quickly to meet these re-quirements; a combined-cycle gas generator can ramp up in 60 minutes. Peak power plants are typically more expensive, but are needed to keep the grid stable.

On the demand side, utilities can encourage consumers to turn off their air conditioners on a hot day, or make deals with households who agree not to use heavy appliances during the week. Demand management seeks to avoid imbalance; supply management to cope with it.

Centralized vs. Distributed Generation - *"Mainframe vs. PC"*
The grid does pretty well in terms of cost, reliability, and providing baseload power. But in other ways, the grid looks clunky and inef-ficient.

For example, today's grid relies to a large degree on sources of gen-eration, like coal and nuclear, that are only economical at large scale; you can't put a coal station on the porch or a nuclear plant in your backyard. Because such plants have to be big, they require a large infrastructure of transmission and distribution (T&D) lines to get that power into homes and businesses. This process is compli-cated. The voltage at the source is generally stepped up to very high levels (higher voltage reduces losses) for delivery to local sub-stations, where the voltage is stepped back down and then distrib-uted through various overhead and underground wires. A significant amount of power is lost in transmission (about 6.5% in the US, according to the Energy Information Administration, and much higher in many developing countries). Another disadvantage of the centralized generation model is that the T&D infrastructure is expensive to maintain and difficult to expand. Due to environmental concerns and "not in my backyard" attitudes, siting new T&D lines can be a nightmare.

Distributed generation inverts the centralized model. Smaller-scale power units—wind, solar, and natural gas all work well—are either at or near where that electricity will be used. Possibilities include gas turbines in the basement of buildings, or solar panels on the roof. Because the juice is not travelling very far, DG places less strain on the T&D infrastructure; in fact, it can often skip the grid altogether. Although the centralized model has the advantage of incumbency—people are used to it—DG is a promising alternative. In fact, it could have a bigger impact on grid design than any technology in the last 100 years.

For a useful analogy, look at computing. In electricity, almost all the generating capacity has historically been centralized, akin to how the early computers were large mainframes. DG is like the personal computer—smaller, nimble, and more individualized.

Is DG growing? Yes. Coal and nuclear power plants are examples of large, centralized facilities. While coal is still growing in China, the United States has clearly shifted away from these large plants. In fact, the only US energy sources to increase capacity over the last 10 years are natural gas and renewables, both of which are smaller more distributed sources. Renewables account for the fastest growing segment, having doubled in size over the past 10-year period, with natural gas increasing more than 50 percent. On a global scale, much of the energy expansion in developing countries is not only distributed, it is off the grid entirely. Global efforts at "rural electrification" often involve small-scale DG sources like solar or biomass;

think of the rural Nicaraguan doctor I described at the beginning of the book using solar to power a refrigerator.

A summary report on DG from Pike Research, *"Distributed Renewable Power Supply to Triple by 2017,"* estimated that worldwide installations of renewable distributed generation will triple in six years, reaching 63.5GW a year in 2017, up from 20.6GW in 2011; four to 10 times faster than global energy growth. Figure 3.2 shows global DG numbers for 2011 and 2012. Europe, led by Germany, has flattened out recently, but only because it has already reached high levels of DG use. The rest of the world is catching up; DG almost doubled in North America from 2010 to 2011, and Asia and the rest of the world showed increases of 19% and 70%, respectively. That is a lot of growth for just one year.

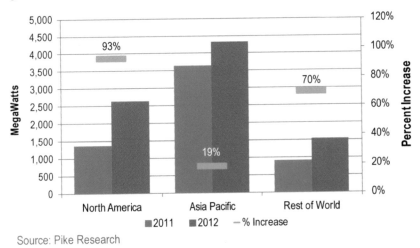

Source: Pike Research

Figure 3.2 – Growth of distributed generation

Distributed generation is growing globally because it brings seven significant benefits:

1. *Improved energy security* - Not only are centralized facilities vulnerable to attack, but the transmission lines to get power to those cities are also vulnerable. Knocking out a single transmission line could shut down New York City or Washington, D.C. While it is possible to guard a nucle-

ar facility, it is impossible to guard the thousands of miles of transmission lines linking that facility to their customers. Distributed generation makes the power system less vulnerable to attack.

2. *A stronger, more robust grid* - Distributed generation makes the grid less vulnerable to unexpected events. Again, an analogy from the tech industry might help. Internet traffic moves through numerous routers and servers, so that if one server goes down, others can pick up the slack, and no one even notices. That's clearly a better approach than using a centralized server hub. In the same way, the availability of enough distributed power could support local needs in the event of a major transmission failure. In a dramatic example, the National Renewable Energy Laboratory (NREL) commissioned a grid stability study after the 2003 Northeast blackout, which was caused by a localized failure in Ohio.[15] The researchers concluded that several hundred megawatts of distributed power sources would have prevented the cascade of events that led to power outages as far away as New York City. Reliance on centralized generation made the grid vulnerable to this massive blackout.

3. *Avoiding T&D losses/costs* - There are costs associated with building and maintaining transmission lines, and with energy losses during distribution. Centralization increases T&D costs. An IEEE report[16] states that "the costs derived from installing, operating, and maintaining the transmission and distribution system have historically comprised about two-thirds of the total costs of producing and delivering electricity to residential-commercial customers, and over one-third of the total costs of supplying electricity to large industrial customers." With DG these costs are minimized because the source is closer to the consumer, minimizing both infrastructure and line-loss costs.

4. *Lower financing risk* – Financing big energy projects can be difficult for a variety of reasons, ranging from permitting and site selection to environmental and tax implications. The smaller a project, the lower the obstacles to finance. And when you get down to the homeowner level, as with rooftop solar, many installers handle the whole permitting and financing project, making it invisible to the customer.

5. *More predictable, stable costs* - A fixed cost of energy is one of the best-known benefits of distributed power systems at the residential and commercial level. For renewables that don't require fuel (like wind, solar, and hydro), there is the known up-front cost of installing the system; after that, operating and maintenance costs are consistent and usually nominal. On the other hand, the cost of electricity purchased from the grid tends to go up over time.

6. *Better access* – More than a billion people around the world do not have access to electricity; moreover, in many locations it is not easy to put together the money and expertise required to run a centralized grid. DG can help to bring power to off- grid communities in developing countries, fueling economic development and helping to alleviate poverty.

7. *Efficient DC grids* – In DG architecture there is less reason to produce AC (alternating current, which is what comes out of wall sockets) because most modern usages of power are DC (what comes out of a battery). The main benefit of AC is derived by long transmission lines (because there are fewer power losses over long lines). In the near future, an all-DC household is possible. This will be more efficient and cheaper, as it will eliminate the need for inverters that convert DC to AC.

These seven benefits make a good case for DG, and I think it is a convincing one. But there is a problem: As energy becomes more distributed, utility companies face new challenges. Utilities are typically built around a transmission and distribution infrastructure, with a single regulated entity controlling the generation and distribution of electricity. But times change...and an industry that has been built on reliability, stability, and centralization for the past century isn't necessarily eager to adapt to that change.

The growth of solar is beginning to disrupt the utility status quo. With increasing numbers of solar panels on American rooftops, and "solar farms" dotting the German countryside, utilities are moving from having a limited number of large power plants that they completely control, to having the grid connected to thousands of distributed energy sources that they do not. This not only creates technical challenges, but threatens revenue and business models as well. As people start to generate their own energy while still connected to the grid, new business models need to evolve to compensate utilities for maintaining the transmission and distribution lines.

~ ~ ~ ~ ~ ~ ~ ~ ~ ~ ~ ~

Renewables, distributed generation, and utility companies are all part of the future of energy...and they're going to need to play nicely with one another to make it work.

Chapter 4

The Case for Solar

Renewable sources of energy will be needed to fill the energy gap while avoiding the harmful consequences of carbon-based fuels. But which ones are best suited to meet future needs? Based on its scalability, accessibility, and compatibility with both the grid and distributed generation, solar is the clear front-runner. Its only disadvantage is that it is the most expensive option — for now. But solar costs are falling fast. Fast enough that for most of the world it is already cost-competitive with other forms of energy, and within five years it should reach grid parity almost everywhere. Solar will be the no-brainer choice for renewable energy worldwide.

To make wide-scale adoption plausible, any energy source needs to fulfill several requirements. The available energy sources and the requirements are shown in Figure 4.1.

	Coal	Natural Gas	Nuclear	Hydro	Geo-thermal	Wind	Solar
Supply	✓	✓	✓	✗	✗	✓	✓
Health	✗	✓	✓	✓	✓	✓	✓
CO2	✗	✓	✓	✓	✓	✓	✓
Intangible	✗	✓	✗	✓	✓	✓	✓
Cost	✓	✓	✓	✓	✓	✓	✓
Summary	✗	✓	✗	✗	✗	✓	✓

Figure 4.1 – Energy sources vs. requirements

First, there is the issue of supply: Is there sufficient supply to provide what we need? (Remember, we're looking at a 46,000GW energy gap.) Second, it needs to be clean, so we must look at how it impacts both human health and the CO_2 levels that impact our planet. Then there are a variety of factors that I include as "intangible": Is it popular, does it work with the grid, does it enable distributed generation, and is it politically favorable in most countries? Finally, there is cost. It needs to be economical; consumers will not sign up for much-bigger electricity bills, and poor people cannot afford to.

Looking at this list no option is perfect, but several choices fall out quickly. Coal is bad on almost every front except cost, and even that is beginning to be challenged by renewables. It is clearly not the energy source of the future. Nuclear is actually good in many areas, but is challenged on the intangibles both globally and locally. There is a global desire for nuclear non-proliferation, so not every country can even get access to the technology. And locally it suffers mightily from the "not in my backyard" syndrome; no one wants a nuclear plant in their town, state, or sometimes, country. Geothermal and hydro are supply-constrained. Geothermal is limited to geographies where the earth's crust is fairly thin. Hydropower needs a river; and many countries have given up on building new dams because of other environmental considerations. Over the last 20 years, growth in both geothermal and hydropower has been flat.

This leaves natural gas, wind and solar as the most viable contenders.

In the effort to reduce coal consumption we will need them all, but they are not all equal. Natural gas is much cleaner than coal, but still produces CO_2 with the attendant climate change consequence. Also, although production has increased, it is still a limited resource which will peak and decline at some time.

Wind is cheap and can be scaled up. But it is weaker than solar in terms of distributed generation, since it's not as economical at smaller scale, and because turbines are usually located where there are good "wind resources" (often far from population centers). Moreover, the timing of wind is the least compatible with the grid because it peaks at night, when demand is low.

Wind's biggest problem has to do with grid stability, meaning the need for constant grid voltage. Wind is intermittent, and as winds gust the sudden surge of electricity upsets the balance of the grid; unbalanced surges result in higher voltages that can trigger a grid shutdown or failure. This is not a hypothetical situation: Germany has been a living laboratory for how high levels of renewable energy affect grid performance. By the end of 2012 Germany had installed more than 30GW of wind power, and has had to deal with power surges caused by these intermittent gusts. It is becoming clear that for large-scale adoption of wind, some grid-level energy storage may be required to absorb the energy spikes.

Now, let's turn to solar. Solar can achieve more scale than any other renewable resource. After all, the sun shines everywhere; almost every country has enough sun to make solar a major contributor. A 2007 study by Nathan Lewis from the California Institute of Technology [17] showed that the sun provides enough energy in *one hour* to supply the world's energy needs for a year.

Solar is also the leader in terms of the potential for distributed generation. In fact, it is the only renewable energy source that is cost-effective at the level of an individual building. It is particularly promising in countries where the grid is weak or non-existent. Solar has the most potential, on a global basis, to bring energy to those who have never had it. As for grid compatibility, let's look at the issues:

> *Reliability* – Solar panels work well, because there are no moving parts. Reliability is not a problem, but intermittency is. Unlike, say, a natural gas plant, solar ebbs and flows; a

cloudy day means little or no output, and of course there is no sun at night. Fortunately, weather predictions allow solar energy forecasts reliable enough to maintain grid stability. As solar energy goes up and down, other energy sources such as natural gas plants can adjust to match the power demand and maintain stability. Again, this is not hypothetical; in high solar sections of California, the reliability has been fine and grid stability has not been a problem. In Germany there has been no grid instability caused by solar, even on days when solar supplied more than half the electricity.

Grid stability and dispatchability - Unlike wind, which is most productive at night, solar is most effective near peak load times, in the late afternoon (there is a small mismatch, as solar peaks usually an hour or two before the energy demand peaks). Solar, however, is not dispatchable; that is, you cannot increase or decrease it at will. So electricity systems will still need peak power plants, such as natural gas, to make up the difference. However, unlike wind, there is less concern about surges or the need for storage to absorb spikes and support grid stability.

Management and control - Although solar excels at distributed generation, utility companies do not necessarily embrace this strength. Utilities are responsible for providing large-scale, reliable energy, and for keeping the grid safe, stable, and balanced. That's easier to with centralized facilities and full control over power generation. With distributed generation, utilities will have to find a way to deal with numerous small power plants that they may neither own nor control. In fact, they are responsible for the entire transmission and distribution system, even to homes that end up paying nothing to the utility because they produce more energy on their rooftops than they consume. So, suffice it to say that the relationship between solar and the utilities is less than harmonious. New policies and business models need to be created to help the two coexist constructively.

In terms of grid compatibility then, solar is not perfect. It does not perform as well as the "always on" renewable sources, such as geo-thermal or hydropower. But it can and does work. Give it an overall rating of "acceptable."

As for the last criteria, cost, this is solar's weak point...at the moment. Solar is generally the most expensive of the renewable options, but its costs have been dropping dramatically. In 1975, solar cost was close to $100/watt. That price fell by half by 1978, in half again by 1981, and then in half again in 1985, 1993, 2006, 2010, and 2012. Depending on the time and place, solar is now approaching the cost of other renewables, and even conventional energy sources. In 2013, a report from Germany's Deutsche Bank stated that solar had reached grid parity (the point at which solar as an energy source can produce power at the same cost as existing elements of the grid) in ten major markets around the world, and was within reach in another 20 in the next few years.[18]

~~~~~~~~~~~~

All-in-all, solar has many plusses but also some challenges with the grid, and a little further to go on cost reduction.

No matter what source is considered, there is no perfect solution to our energy needs. That is why many respected agencies, such as the International Energy Agency and the US Energy Information Administration, and even some "Big Oil" stalwarts like Shell Oil, have come to the same conclusion. The following is a quote from Peter Voser, who was Shell's CEO until December, 2013:

> *Our energy consumption is on a scale so massive and demand is growing so quickly that we will need to aggressively pursue all sources of energy for decades to come just to keep up," said Voser, whose company has 20,000 US employees. "When President Obama says America's energy policy should be all of the above he is absolutely right, in our opinion.*

The "all-of-the-above" strategy is the only realistic solution. And "all" includes renewables; a lot of them. Of all the choices in renewable energy, solar comes closest to meeting all the criteria. If it closes the price gap, it will be the outright winner in just about every category. In the next chapter, I will discuss the economics of energy in general and solar in particular, with a detailed look at the all-important cost issue.

Chapter 5

Solar Economics 101
"Cost is Everything"

> *Electricity is a commodity. Few of us think about how it gets into the house, but many of us care about how much it costs. Indeed, almost every chapter in the rest of the book will include some discussion of costs.*
>
> *To establish a common language, in this chapter I define certain terms, such as the levelized cost of energy, solar yield, and total installed cost; and then explain how they apply to solar. If you are familiar with these concepts, you can probably skip this chapter. If not, please slog through it. Understanding the cost dynamics is critical to understanding solar.*

The costs of producing electricity vary depending on local conditions. Americans who live close to a hydroelectric dam, such as in the Pacific Northwest, pay less because hydropower is so cheap to produce. The cost structure also varies. Solar, for example, requires more upfront investment, but has no fuel costs (since sunlight is free) and low maintenance costs (because there are no moving parts). Wind is similar in that there are upfront costs and the fuel is free, but the maintenance costs are higher than for solar. With the drastic drop in solar and wind costs at the utility scale, a new coal plant now costs more to build and has higher operating costs because of the maintenance and the ongoing fuel expense.[19] This of course has contributed to coal's decline in many countries. Still, a coal plant has the advantage of being able to run almost 24X7, whereas a solar plant only operates when the sun is shining. Natural gas plants are cheaper to build than solar, but have more on-going costs because of fuel. Each situation is different, and all these factors need to be considered in the cost equation.

Before we look at this "cost equation," we first need to understand how electrical power is measured. It is based on how much power is used (watts), and how long it is used (hours). The industry uses the metric of 1,000 watt-hours, or kilowatt-hours (kWh). The method used to calculate the cost to produce a kWh of energy is the "levelized cost of electricity," or LCOE.

LCOE

The LCOE is the gold standard for cost calculations because it includes all the costs, from financing to construction to operations and maintenance, and the total output of a power plant over time. It therefore provides an apples-to-apples comparison of the cost to generate electricity from any available source, whether conventional or renewable. Calculating LCOE requires a complex formula: total installed costs (see below), plus operating and maintenance costs, divided by the total energy output (given in solar yield). To put it another way, LCOE comes down to two questions: How much will it cost to build and operate the power plant over its lifetime? And how much electricity will the plant produce over its lifetime?

One shortcoming of LCOE is that it generally ignores the cost of transmission and distribution (T&D), which can be up to two-thirds of the total. Distributed generation, of course, uses less T&D infrastructure, which improves its economic profile compared to conventional sources. Recently, there have been efforts to estimate the value of this avoidance of T&D cost. Austin Energy, a Texas utility, is experimenting with a value-of-solar-tariff (VOST) algorithm, which takes into account avoided T&D costs. Something like VOST better captures the benefits of DG, including solar, while LCOE puts DG at a disadvantage. Even though LCOE is king, because of the complexity of calculating it, people often use other means of comparing costs.

Total Installed Cost

With wind, solar, hydroelectric, and geothermal, operating costs are low because the fuel is free; therefore, total installed costs (TIC), measured in dollars-per-watt ($/watt) of power produced, dominate the cost end of the LCOE equation.

Let's take a look at solar's TIC. The hardware costs include not just the solar panels themselves, but all the mounting and electrical hardware to connect them together and to the grid (if it's being used). Then there are "soft costs," such as labor, permitting, financing, overhead, and profit margins.

Historically, the solar panel itself has been the most expensive feature of a solar system. In recent years, though, as panel costs have plummeted, all the other costs, or the balance of system (BOS) costs, have exceeded panel costs. Hardware is getting cheaper, and soft costs have become the major industry concern. For example, interest rates on loans to install solar projects are relatively high, because investors consider solar risky; these are likely to come down though, due to very low default rates on solar loans. That would help the overall finances considerably. Figure 5.1 shows how LCOE varies from $0.25/kWh up to $0.90/kWh based only on changes in financial terms.[20] (Note: 2014 results are almost half the estimates made in this study, as install costs have gone from $5/watt to closer to $2.50/watt.)

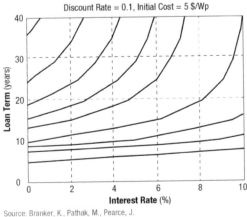

Source: Branker, K., Pathak, M., Pearce, J.

Figure 5.1 - LCOE vs. loan term and interest rate

Solar Yield

TIC is about the first part of the LCOE equation, costs. "Solar yield" is about the second—the amount of electricity produced at a given location over time. Solar yield is measured in kilowatt-hours per installed kilowatt. In other words, for each kilowatt of solar powered installed, how many kilowatt-hours of energy are produced in a year? Numbers range from 700 to 2,800. This is a wide spectrum, but it makes sense when you compare a sunlight-challenged northern city like Seattle to a desert-like southern city such as Phoenix.

Four factors determine solar yield:

Efficiency – This is probably the most talked-about metric in the solar industry. Efficiency is the ratio of actual energy generated from a solar panel vs. the total amount of energy in the sunlight that hits it. It is measured in percent. In effect, you shine light on a solar panel and measure the electricity output. The accepted standard sunlight intensity is 1000 watts per square meter. So if a one-square-meter panel produced 1000 watts, that would be 100% efficiency. The norm, however, is more like 100-200 watts, or 10% to 20% efficiency. A higher-efficiency panel obvious means more power output, everything else being equal.

Solar Insolation - The amount of sunlight that arrives is called the "solar insolation," and is measured in a term called "direct normal irradiance" (DNI). This has been mapped out everywhere in the world. The latitude of the solar plant's location determines the number of hours of sunlight it receives each day. Weather patterns, such as the amount of cloud cover, are also important; Portland gets a lot less sunlight than Palm Springs.

Ideally, the panel should be tilted toward the sun at the same angle as the latitude of that location. For example, in the US, panels should be tilted roughly 22 degrees up from flat and facing south. Sometimes people install panels at whatever angle the roof provides, and the closest direction to south. This is okay, but it will reduce the solar yield.

Diffuse and low-light performance - Diffuse light is light that is scattered by molecules in the atmosphere, mostly water vapor in clouds. Low light is experienced around sunrise and sunset when the sun's rays are at a very low angle, and is not absorbed as well by solar panels. Direct sunlight is best to produce the maximum amount of solar energy. But since not every day is sunny, systems that more efficiently convert diffuse or low light into power have a higher solar yield.

Degradation mechanisms - Solar panels slowly degrade, and these losses add up over a typical 25-year lifespan. Each material used to make solar cells has a different rate of degradation. System designers should chose solar panels with the "best" degradation characteristics for their specific conditions. For example, one degradation mechanism is thermal—the power output goes down as the panel gets hotter. This can be a problem, since panels are often sited in the hot sunlight. So a desert location might use a different system, one with better thermal degradation characteristics, than one designed for the cold of Alaska. Typically, system designers allow 0.5-1% degradation per year over the life of the system.

Because solar yield varies so widely, LCOE can be difficult to calculate. That is why many industry analysts have migrated to TIC as their preferred cost metric; after all, the up-front costs reflect most of the money spent. (This book will mostly use the TIC $/watt metric.)

~ ~ ~ ~ ~ ~ ~ ~ ~ ~ ~

This chapter has explained the basics of solar economics; the next one looks at the industry's structure. Then we will move onto to most interesting part of the story: how solar has passed the tipping point, and what we can expect in Solar 2.0.

Chapter 6

The Solar Ecosystem

Understanding how the solar industry works is critical to understanding Solar 2.0 and future chapters. The industry structure includes three main segments: upstream, panel (or module) manufacturing, and downstream. "Upstream" refers to the materials and equipment needed to make solar panels. Within the "panel manufacturing" segment, there are different business models for silicon-based companies versus those focused on thin film technology. "Downstream" refers to everything else required to install and operate solar-powered systems, including BOS (balance of system) components, service, and finance. Finally, solar power systems are delivered to either residential, commercial or utility-scale "end users". (This material may be familiar to industry veterans — feel free to skip or skim.)

Those solar panels on your neighbor's roof may look simple. They are, but getting them there isn't. It requires a complex process of negotiating a supply chain, building products, and getting them installed. An entire ecosystem is involved.

As shown in Figure 6.1, there are three distinct but connected segments within the solar industry: 1) the upstream market is everything that goes into building solar panels; 2) then comes the center of the industry, the panels; 3) the downstream market is everything needed after the panels to build a solar power plant, and get the electricity to the end users. In this chapter, we describe the different elements that make up each segment.

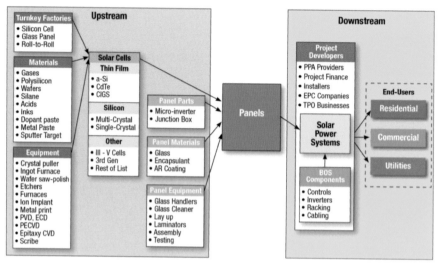

Figure 6.1 – Solar ecosystem

Upstream

In the past, the upstream segment has accounted for the majority of solar costs. That is no longer the case. Every link in the upstream supply chain has been analyzed and optimized; at this point, costs cannot go much lower.

Figure 6.2 shows the supply chain for silicon solar panels. The thin film supply chain is simpler as it leaves out the polysilicon, ingot, and wafering steps, going directly from materials and equipment to cell processing.

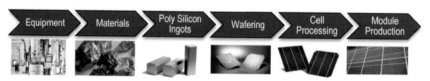

Figure 6.2 – Silicon upstream supply chain

The supply chain reflects the manufacturing process flow: Equipment is used to make materials, like polysilicon, which is used to make ingots, which are used to make wafers, which are used to make cells, which are put into modules. There is one anomaly: Equipment is used not only at the beginning, but in every step of the process. The upstream segment ends once you have everything needed to make a module.

Equipment: Equipment impacts both module costs and performance. As of early 2014, the industry is just coming out of a difficult part of the cycle in terms of equipment. Although much of the industry overcapacity has been absorbed, there are very few capacity additions. As a result, little new equipment has been needed to increase production. Solar equipment suppliers tried to diversify into other markets, but low sales have led to bankruptcies and consolidation. But the bottom of the cycle may have passed, as some analysts forecast increased equipment sales in 2015.

Materials: Unlike equipment, which is a customized, low-volume, high-priced business, the materials segment is more like a commodity, featuring low prices and high volumes. There is little opportunity left for cost reduction. There are numerous markets for the materials used in solar, so most of the top suppliers are already operating at scale. The key materials for silicon solar are polysilicon (poly), glass, and silver.

Poly is important because historically it is the most expensive material in the silicon solar cell. It is a mature industry with a long history. Making poly is a capital- and energy-intensive process with plants costing $500 million to $1 billion and taking a year or more to build. The market for poly is shared between solar and semiconductors, but for most of history the demand was dominated by semiconductor requirements. As late as 2000, only 10% of poly was used for solar cells. The poly industry exhibited a steady growth of 10% to 15% a year, and prices hovered around $30/kg.

Then things changed. Solar demand took off, and by 2008 solar accounted for over 50% of poly demand. Combined with the existing semiconductor requirements, the demand for poly soared. The sleepy poly industry became the center of the solar world's attention. The solar-driven demand spike drove prices up to as much as $400/kg (Fig. 6.3). Investors piled in, supply surged — then prices collapsed. In fact, due to overcapacity, by July of 2013 poly was selling below manufacturing cost. This price excursion created havoc in the industry, but poly prices have already recovered and will likely settle in at $22 to $24/kg, which is 20% less than the historic level.

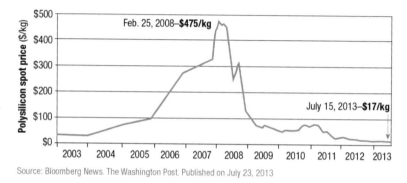

Source: Bloomberg News. The Washington Post. Published on July 23, 2013

Figure 6.3 – Polysilicon spot prices

Ingots: Polysilicon is melted and then solidified into boules, or blocks of crystalline silicon called ingots. There are two types of ingots: single crystal and multi-crystalline silicon. Single crystal (sc-Si) silicon, is essentially a very large, single crystal of silicon. This is the highest-quality and most expensive form of silicon, and delivers about 20% higher efficiency compared to the second type, multi-crystalline (mc-Si) silicon. Instead of one crystal, mc-Si has many hundreds or even thousands of crystals in a single wafer. It is cheaper than sc-Si, and even though it is not as efficient, it delivers more power for the same dollar. Mc-Si solar cells are the mainstay of the industry, accounting for over half of the market. The term "c-Si," for crystalline silicon, refers to both types. You'll see these abbreviations used throughout the book.

Wafering: Wafering is the process of sawing c-Si ingots into thin wafers. Mc-Si blocks are rectangular and are cut into square wafers;

sc-Si boules are a like salami sausage and are sliced into round wafers. The wafers are then cleaned and polished.

Cell processing: The final step in the upstream supply chain is to convert the wafers into solar cells. A series of well-understood thermal and chemical processes alters the silicon and makes it photovoltaic, meaning that it can then convert sunlight into electricity. When building a solar cell factory, one can buy the process equipment from vendors like Applied Materials and Centrotherm, buy the wafers from MEMC or Wacker, and run the established processes. As a result, there is little differentiation between cell processing companies. Since there are few barriers to entry, anyone with the money and desire can get into the solar cell business. The only way to differentiate is on price, and that fact has eroded margins and pushed many companies out of business.

Solar Panels (or Modules)

The upstream supply chain provides the materials and equipment to make solar panels. There are two types of panels: silicon and thin film. Both convert sunlight into electricity, but the manufacturing processes are so different that they are considered different market segments.

Silicon solar: Electronically, a silicon solar cell acts like a "diode," which is a semiconductor device. That is why most of the technology, equipment, and materials for solar over the last 30 years have come out of the semiconductor industry. The semi industry is highly developed with a robust supply chain stuffed with multi-billion dollar companies. This benefits silicon solar.

There are only a few simple processing steps to assemble solar cells into solar panels, and there is little differentiation among suppliers. The resulting low margins have driven numerous solar cell and solar panel manufacturers to merge, creating a high degree of vertical integration. China, through aggressive expansion, has come to dominate this sector.

Thin film solar: Thin film is different from silicon because the materials and equipment technologies are relatively immature and extremely specialized. In fact, nearly every company creates its own absorber—the material that absorbs and converts light into electricity—using customized processes and equipment. Although thin film technology was the first to go below $1/watt cost, lack of standardization has hampered further costs reductions, and few thin film technologies have reached significant scale.

The exception is First Solar, which has achieved both scale and low costs. Thin film market share increased in the early to mid-2000s as First Solar dominated the market, becoming the first company to reach 1GW capacity and beating all silicon suppliers. Investors, excited about First Solar's success, poured in $6 billion to fund over 100 thin film startups. But when China's silicon companies flooded the market with cheap solar panels, many investors gave up and most thin film companies failed. This dramatic rise and fall of thin film is shown in Figure 6.4. Thin film has not recovered yet.[21]

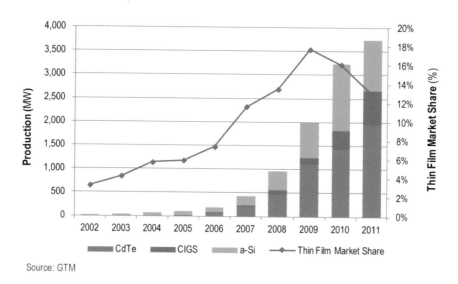

Source: GTM

Figure 6.4 – Thin film production and market share

Even so, thin film has a chance. First Solar is a tremendous success story; in early 2014, it was the most profitable solar panel manufacturer in the world. But the industry needs to do better: lowering costs, improving management, and devising new business models and strategies.

Downstream

The cost of solar panels has fallen so far that they are no longer the main cost factor; the balance of system (BOS) costs matter more. These BOS costs define the downstream segment; everything required for deployment of solar panels in the market. This segment, shown in Figure 6.5, includes BOS hardware components; engineering, procurement and construction (EPC); finance companies; and developers.

Figure 6.5 – Downstream segment

The downstream market is not only seeing innovation in technology, but also "market" innovations involving new business models to address soft costs, such as finance, labor, permitting, and overhead. Downstream is truly a complex, exciting, and target-rich environment. Let's take a look at each downstream component.

BOS components: This covers everything needed to build a solar power system, other than the panel itself.

> *Inverters* convert the DC power coming out of solar panels into the AC power that runs homes and offices. They are purely electrical devices, and the major players tend to be large, international electrical firms. The key points about inverters are reliability and system architecture, specifically where to do the DC-to-AC conversion. This can be done on each solar panel (by micro-inverters of the kind Enphase makes), or after a string of solar panels is wired together (string inverters), or just in one spot

after all the solar panels are interconnected (central inverters). Inverters are more expensive and less reliable than other BOS components. There is definitely room for innovation on inverter cost and quality.

Hardware refers to the mechanical and electrical hardware needed to mount solar panels, and then electrically connect them together so we can use the electricity. These are largely commodity items supplied by large distributors; they have little technical content.

Soft costs are now the single biggest cost factor in solar, but there is no physical component and therefore no supplier base. Many new business models are being devised to reduce transaction costs, permitting, labor, and financing cost.

EPC: stands for engineering, procurement, and construction. Every solar system needs to be designed and built; this is the EPC function. For residential projects, the EPC company might be an electrical engineer with a specialty in solar, or a homeowner could take on the installation as a do-it-yourself (DIY) project. EPC is more critical in commercial and utility-scale projects. The commercial EPC market has a regional mom-and-pop flavor, similar to the roofing, electrical, or plumbing trades. There is no strong national firm, but a number of strong regional players. Utility-scale projects typically use multibillion-dollar, global EPC firms like Bechtel and CM2Hill that have been building huge projects like pipelines, dams, and power plants for years.

Solar finance: Financing is crucial because almost all the costs of solar come at the time of installation. In effect, the customer is paying for 25 years of solar energy (the typical lifetime of a system) all at once, and that can make the price tag look daunting. Even if the economics say it is a good long-term investment, residential costs can be $10,000 to $20,000, a sum that many people don't have available. This upfront loading can slow the adoption rate of solar.

The development of solar leases, which are growing two to three times as fast as the industry as a whole, address this problem. Known as third-party operators (TPO), leasing companies handle the financing of the solar power system so that consumers do not have to come up with such a big upfront payment. TPO companies also need financing as they cannot cover the upfront costs of hundreds of megawatts of installations themselves. However, they are beginning to get the capital they need (Fig. 6.6). SolarCity alone raised more than $1 billion for financing solar installations.

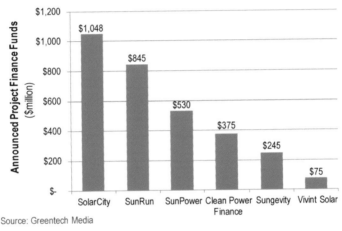

Source: Greentech Media

Figure 6.6 - Project finance raised by TPO companies, 2013

The big banks, including Bank of America, Goldman, and Citicorp, are getting involved. And it is not just in the United States: Japanese banks announced in early 2013 that they would be increasing domestic solar investments as much as eight-fold, up to $19 billion.[22] Apparently, a lot of people whose business is money see that there is opportunity in solar.

A final note on financing: To get a bank loan for a solar project, the components being used must be "bankable," defined as having demonstrated reliability over time, or having a long track record of success. EPC companies cannot just select the best components when designing a solar system; the bank must also approve the design. Such oversight is a symptom of an immature industry. As tests, data, and quality controls improve, banks should step back.

Developers/third-party operators (TPOs): On many solar projects, especially large ones, someone needs to coordinate all the work: Design the system, get the permits, buy the components, build the system, then maintain and operate it. On a small system this might be the owner, but on large installations, each function is done by a different entity. A developer coordinates all this work: EPC, finance, permitting, installation, and operations. With the TPO model, this service is now available even for the homeowner. A company will finance, design, install, and operate a residential solar system; all the homeowner has to do is sign a contract. This lowers the cost of entry, in both financial and logistical terms. This new TPO category of developers is growing fast, with exciting companies like Vivint, SunRun, Sun Edison, and the leader, SolarCity.

End Users

There are three kinds of end users for electricity: residential, commercial, and utility-scale. The market share of each segment varies by country, and even within a country will vary by region or state. Germany, the market leader for years, is predominantly a residential and commercial market. The US, on the other hand, has recently been dominated by utility-scale installations, with that segment making up more than half of recent installations (Fig. 6.7).

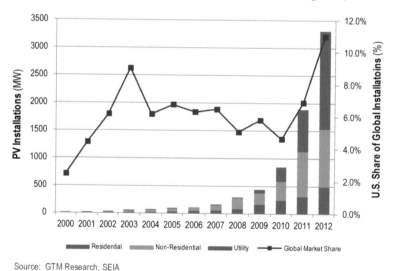

Source: GTM Research, SEIA

Figure 6.7 – US market growth vs. average system price

But with the pipeline for utility scale narrowing, and the TPO model expanding to the residential and commercial markets, this trend is set to reverse. Let's look at each segment in a little detail.

Residential is just that—solar panels on single-family or apartment rooftops that provide some or all of that household's power. The systems are small, ranging from of 1,000 to 10,000 watts (about 4,000 on average). Each system needs to be designed for a specific roof, and residential roofs tend to vary quite a bit. To accelerate adoption, solar companies that are focused on residential markets are trying to make it easier by providing the complete end-to-end service of a solar developer. Such TPO arrangements account for the majority of new residential installations.

Commercial systems are larger than residential ones (20 kW to 1,000 kW), but smaller than utility scale. Typical commercial installations might be on the roofs of warehouses, big box stores, and other industrial buildings. The design is still customized, but there is a degree of standardization because most of these roofs are flat. There are millions of these roofs just sitting there, waiting to be put to work. The tricky part of this market is that the owner of the building may not be paying for the electricity. Businesses that lease their premises might be motivated to install solar, but they can't because they don't own the building. Meanwhile, owners are unmotivated to make such investments because they aren't paying the utilities. This is why most commercial installations are on owner-occupied buildings—big box brand name stores in particular.

Utility-scale refers to massive arrays of solar panels that can cover several square miles and generate 500MW to 1GW of electricity (the latter being enough to power 750,000 American homes). The complexity of utility-scale solar is not in the hardware, but in all the "soft" stuff. Plants can cost $2 billion to build, so banks are heavily involved. An environmental permit is usually required, so groups like the Sierra Club and an array of government agencies might get involved. The customer is the utility company, which is government regulated, so there is a ton of paperwork, and politicians might get involved as well. It is not surprising that utility projects can take five years to complete.

Utility-scale solar was all the rage in the US in 2011 and 2012 due to RPSs Renewable Portfolio Standards (RPS); these are regulations that require utilities to provide a certain percentage of their energy from renewable sources. Most other countries have favored more distributed generation sources, like residential and commercial-scale applications. Globally, the trend is toward the simplicity of smaller-scale projects that can be located on-site or close to the end user. I think this trend toward distributed generation will continue, but utility-scale will still play an important and significant role.

~ ~ ~ ~ ~ ~ ~ ~ ~ ~ ~ ~

Now that we have established a foundation about solar economics and industry structure, we'll move on with the story.

Chapter 7

The History of Solar

This was one of the most interesting chapters to research and write — I hope you'll find it interesting too! So far, the growth of solar has been almost totally driven by technology and policy. Government policy has stimulated both supply (by providing factory loans) and demand (through various subsidies), which in turn has driven costs down. In fact, the phases of solar history are all about different levels of cost. Initially solar was more of a "hobbyist" activity and technical curiosity, where cost was mostly irrelevant. Solar 1.0 saw governments getting involved to drive both R&D and demand, and costs dropped due to technology innovation and economies of scale. All this leads us to a tipping point (which we'll look at in the next chapter), and the beginning of a long-term phase of sustainable, economically driven growth in solar.

For most of its history, solar was mostly just a technical curiosity. At a whopping $100/watt, it had almost no commercial value. In this "hobbyist" phase, most people who were involved with solar either lived off the grid and/or were just technology geeks. There was almost no industry interest due to the poor economics, nor was there much government attention, other than as a power source for satellites and a few other specialized, cost-insensitive applications.

Enter Solar 1.0

All that changed with the oil crisis of the 1970s, the beginning of the period I call Solar 1.0. Countries began looking for alternative sources of energy. In the US, concerns about energy independence led to the creation of the Department of Energy and the Solar Energy Research Institute (SERI, which later became NREL). From 1978-1981 an average of $6.3 billion per year was spent on energy R&D,[23] and renewable energy got a nice piece of that; almost $1.3 billion a year for four years.

Then the Reagan administration cut renewable energy investments by 80%, and many solar researchers had to find new jobs. But the R&D investment was starting to pay off: The cost of solar came down an unbelievable 50X over two decades (Fig. 7.1). Nevertheless, solar was still not cost-competitive with conventional sources of energy.

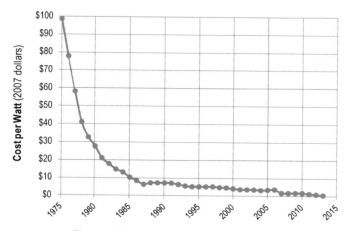

Figure 7.1 – Module cost, 1975-2013

Most of the early days of solar revolved around silicon technology. Then in the late 1980s, NREL researchers began to investigate thin film materials like amorphous silicon (a-Si), cadmium telluride (CdTe) and copper-indium-gallium-selenide (CIGS). These materials require less thickness to absorb the same sunlight as traditional silicon. The thinking was that these thinner materials could lower costs even more.

Enter First Solar: In 1990, Harold McMaster started Solar Cells, Inc. (SCI) to focus on one of these thin film materials, CdTe. As SCI began producing full-sized panels, Walmart began looking at the possibility of using solar in their stores. In 1999, the investment arm of Walmart, True North Partners, purchased SCI and rebranded it as First Solar. The rest is history...at least in the solar industry.

First Solar began shipping its first-generation CdTe solar panels in 2002. Five years later, they expanded their US production facility and became the global low-cost leader, with a cost of just $1.08/watt in 2008. A year later that was down to $0.89/watt; sales rose, and First Solar became the world's leading solar-panel manufacturer.

Enter Germany: Just as US government policy had driven solar panel prices down, policy in Germany did the same almost 20 years later. In the late 1990s, Germany's Social Democrats formed a coalition with the Green party. A condition of the latter's support was to promote green energy, including solar. As a result, Germany came up with a system, the Feed-in-Tariff (FIT), which required utilities to buy renewable energy from basically anyone who produced it, at a government-set (and attractive) price. The idea was to provide an incentive for people to install wind and solar power systems—and it worked. Almost too well: the government has reduced the FIT several times to manage the amount of renewable energy coming online. Germany became the wind and solar capital of the world, with more than half the world's installations in 2010 (Fig. 7.2).

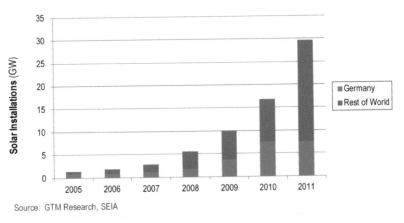

Source: GTM Research, SEIA

Figure 7.2 – Solar installations, Germany vs. rest of world

To put this into perspective, the total amount of solar capacity installed globally in 2005 was 1.4GW. Germany has averaged more than double that over this whole period, accounting for more than 40% of all solar installed in the world. Government policy once again drove prices down and volumes up — and China got interested.

Enter China: China had been fueling its amazing economic growth chiefly by burning its abundant coal reserves, resulting in alarming levels of pollution. China was not going to give up on growth, but by the early 2000s, its filthy air was becoming an international embarrassment, and a national health crisis. So the government began investing in clean tech, particularly solar, in a big way. China's entry into the market completely transformed it. Chinese manufacturers added a huge amount of capacity, chiefly in silicon solar.

Around this same time, there were stirrings around the world about a new concern: climate change. Was the rising temperature of the earth a natural phenomenon, or was it related to all of the CO_2 we have released into the atmosphere by burning carbon-based fuels? And was this rising temperature a sign of irreversible climate change? Working with Jeff Skoll, former US Vice President Al Gore created the movie *"The Inconvenient Truth,"* and the attention this created helped intensify the drive to reduce carbon-based energy sources. In progressive states like California, policy action was taken to stimulate demand for renewable energy, including solar.

The Perfect Storm

Now we had Germany creating huge demand, the US investing in technology that could lower costs, First Solar driving costs down to exciting levels, China needing energy to fuel its growth, and climate change driving people into pushing for more renewable energy. China decided that renewable energy was a cleaner way to satisfy its own appetite for energy, and that solar could create jobs and be a lucrative export product.

Once again, government policy came into play. China's national government provided huge incentives for cities to build solar factories, in essence subsidizing supply. They selected the more mature technology, silicon, because it would be easiest to ramp quickly. Huge manufacturing capacity came online in China, adding to the hefty capacity already in pace in Germany, and costs dove down (Fig 7.3).

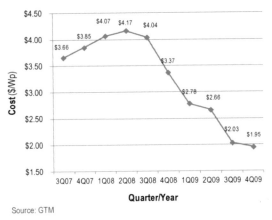

Source: GTM

Figure 7.3 – Sharp decline in Chinese module price

By the end of 2009, silicon solar was chasing First Solar's thin film solar on cost. And that was just the beginning. Silicon solar prices were cut in half again over the next year, reaching ~ $1.00/watt in 2010. And it didn't stop there: By the end of 2012, silicon panels could be found priced at $0.60/watt, below the manufacturing cost of about $0.80/watt. Overcapacity had driven panel sales prices below manufacturing costs.

~ ~ ~ ~ ~ ~ ~ ~ ~ ~ ~ ~

The combination of China's capacity and Germany's demand created the conditions for a new era in solar. In this new phase, prices are low enough to be economically competitive with conventional energy sources. In the next chapter we'll look in more detail at how we have passed a critical tipping point on cost.

Chapter 8

The Tipping Point

The tipping point is when it becomes inevitable that solar will be cost-competitive with conventional energy sources. The technical term for this cost-competitiveness is "grid parity." There is a well-established learning curve that predicts lower costs based on higher volume. A virtuous cycle is formed when higher volume leads to lower costs; which leads to even higher volumes, which in turn leads to even lowers costs. Solar is already at grid parity in a number of markets. As it spreads across the globe, we are beginning to see the emergence of an economically-driven solar industry of the future, as opposed to the subsidy-driven industry of the past.

There is no single price point at which solar reaches grid parity; at any given location, grid parity is the point where the local cost of solar is equal to the local cost of grid electricity. It is easier to reach grid parity in areas where the price of electricity is high, but in places like Idaho (where electricity costs only \$0.06/kWh), it's more of a challenge. The bottom line is that lower solar costs translate into higher rates of grid parity.

I have found no data that shows grid parity everywhere on the globe, but Figure 8.1 is a conceptual graph of what that might look like. At zero cost, solar would of course be at grid parity every-where. But at some high price, like over \$0.50/kWh, solar would be only at grid party at a very few high-cost electricity locations.

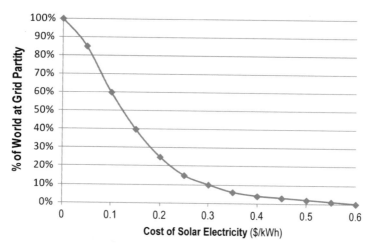

Figure 8.1 – Conceptual grid parity chart

The tipping point for solar is when it becomes inevitable that we will achieve grid parity in most markets. This is tricky to call, because "inevitability" is a slippery concept. But we can make an intelligent prediction by understanding what's driving the cost reductions and the likelihood of the trends continuing. Once we've done the heavy lifting and pushed our way to the tipping point, the forces that got us there will inevitably carry us to the finish line — defined by the US Department of Energy as a total installed cost (TIC) of $1/watt.

SunShot Goal - $1/watt = Grid Parity

In 2010, the US Department of Energy developed an initiative to reach a target of $1/watt TIC by 2020. Reaching this goal would bring solar to grid parity not just in the US, but in most markets of the world. This goal (known as SunShot) not only looked ambitious, as costs were close to $4/watt at the time, but so ambitious as to verge on the ridiculous. At the time, that was the opinion of SunPower's Dick Swanson, who is widely acclaimed to be one of the fathers of the solar industry. By 2012, Swanson had changed his mind. Giving a presentation to the IEEE Silicon Valley PV group at the Palo Alto Research Center, he argued that in fact the SunShot goals would be reached well before 2020. Here is how Swanson went from skeptic to believer.

The TIC is the sum of four parts: the inverter, BOS (balance-of-system) hardware, BOS soft costs, and the module (or panel). Figure 8.2 shows the actual US figures for 2010, and the projections to meet the 2020 SunShot goals.

Component	Residential		Commercial		Utility	
	2010	2020	2010	2020	2010	2020
Module	$2.10	$0.54	$2.00	$0.52	$1.90	$0.50
Inverter	$0.40	$0.12	$0.35	$0.11	$0.30	$0.10
BOS Hardware	$0.50	$0.19	$0.70	$0.18	$0.40	$0.18
BOS Soft Costs	$2.70	$0.65	$1.55	$0.44	$1.20	$0.22
Total	$5.70	$1.50	$4.60	$1.25	$3.80	$1.00

Figure 8.2 – Actual 2010 US costs and 2020 SunShot goals ($/watt)

These goals are $1.00/watt for utility scale, $1.25 for commercial and $1.50 for residential installations. The case will be made that costs are declining now, and we are on target to hit the SunShot goals for each segment.

Inverter: Of the hardware components, aside from the panel itself, inverters are the next most expensive. In Dick Swanson's presentation, he indicated that high-quality inverters are already close to the 2020 goals. He suggested that if you were willing to take the risk and buy inverters from 2nd- or 3rd-tier suppliers in China, you could already get inverters at less than $0.10/watt, below the SunShot goals. If they are being sold in China today at the target goal, it seems clear this component will not be a problem in achieving grid parity.

BOS Hardware: These electrical and mechanical components are low-tech with little differentiation, and as volumes increase, these commodity products will come down in price. Leading manufacturers on both the mechanical and electrical sides have stated they will meet or exceed the SunShot goals.

BOS Soft Costs: These vary widely, depending on local labor costs, interest rates, policy, building codes, and bureaucratic red tape. With hard costs declining rapidly, soft costs have become a larger percentage of TIC, and the center of attention for future cost reduction. The good news is that everyone is working on it. Businesses are springing up with solutions to reduce both financing and installation costs. Many governments have revised policies to make solar installation easier. The US government, through the DOE, has initiated a soft cost program, and Germany has done even more. In fact, the experience of Germany (Fig. 8.3) provides the best evidence that the DOE's SunShot soft cost forecast is doable.

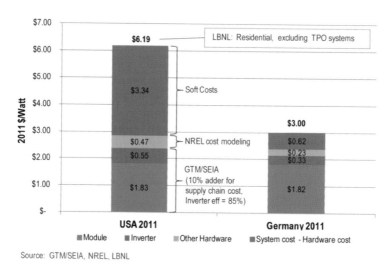

Figure 8.3 – Installed cost breakdown, US vs. Germany ($/watt)

With close to 500% more solar installed than the US, Germany is farther down the learning curve and they have made admirable progress. Their soft costs dropped to $0.62/watt for residential installations in 2011, even lower than the SunShot goal. Other countries, including the US, are still far behind, but Germany has shown what can be done.

Modules: By the end of 2012, module costs were already close to $0.65/watt due to substantial over-capacity. But the price rebounded somewhat in 2013 and has stabilized at around $0.70/watt. Roughly a 30% cost reduction is needed to meet the SunShot goals.

When prices go down in a price-driven commodity market, demand usually goes up. Shipments increase to meet demand, and costs of production decline as manufacturers take advantage of the economics of scale. A virtuous cycle forms: volume increases lead to cost reductions, which lead to further volume increases, which lead to further cost reductions. This is the famous cost learning curve for solar, shown in Figure 8.4. The end result is that every doubling of cumulative volume has led to an 18% drop in costs.

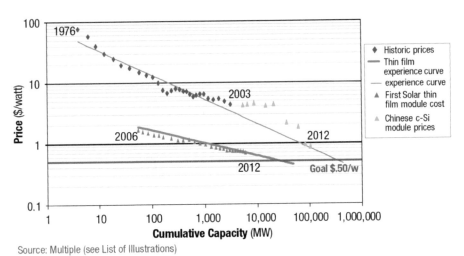

Source: Multiple (see List of Illustrations)

Figure 8.4 – Learning curve, high module volumes drive lower costs

There are a few interesting stories embedded in this chart. The first is that prices fell by 90% as volumes went up 5,000 times. Another is the "hump" between 2005 and 2009. This was a blip caused when solar demand for polysilicon surpassed that of the entire semiconductor industry. Prices of poly rose by as much as 2,000% as demand far outstripped supply. The third story is China. It entered the market in the mid-2000s with a vengeance, and the industry hasn't been the same since (see box below). China added so much capacity that panel prices fell to historic lows, bringing prices back in line with the historic learning curve.

Although no one has yet reached the $0.50/watt target (the red line in Fig 8.4), the trajectory is very clear. Both c-Si and thin film module technologies are projected to cross the goal line in the not-too-distant future. In fact, an industry report by Greentech Media has manufacturing costs at $0.37/watt by 2017.[24] With an industry standard 20% gross margin, that would put the selling price of panels below the $0.50/watt goal four years ahead of schedule.

The China Syndrome

Through generous government subsidies and preferential access to finance, China entered the market in the mid-2000s, and by the end of the decade became the major factor in both the poly and the solar industries. China added more than 30GW of capacity, driving prices to unheard-of lows.[25] Panel prices dropped from $4 to $2 in just 18 months; by the end of 2013, the average selling price (ASP) was down to $0.65/watt. This further stimulated demand and accelerated the virtuous cycle.

Not only did this cost reduction accelerate grid parity for solar, but it established China's market leadership. In 2004, no Chinese firm ranked in the top-ten solar companies; in 2012, six did (Fig. 8.5).

Rank	2004	2008	2012	Legend
1	Sharp	Q-Cells	Yingli Green Energy	Japan
2	Kyocera	First Solar	First Solar	USA
3	BP Solar	Suntech	Suntech	Europe
4	Q-Cells	Sharp	Trina Solar	Taiwan
5	Mitsubishi Electric	Motech	Canadian Solar	China
6	Shell Solar	Kyocera	Sharp	Korea
7	Sanyo	JA Solar	Jinko	
8	RWE Schott Solar	Yingli Green Energy	JA Solar	
9	Isofoton	Gintech Energy	SunPower	
10	Motech	Solar World	Hanwha Solar One	

Figure 8.5 - Top 10 global solar-panel manufacturers (by volume)

Installed Cost Forecast

Bringing it all together, there is clear momentum in cost reduction. Some elements, such as modules and inverters, are already close to SunShot goals. Using both internal and external data, Figure 8.6 provides my forecast for total installed costs at the residential level, and can be used to estimate when grid parity will be reached in various places around the world.

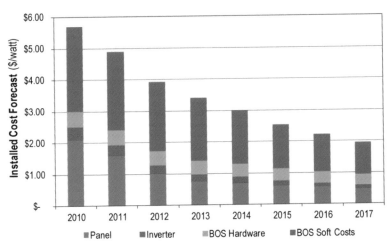

Figure 8.6 – Forecast of TIC, total installed cost (author forecast)

Grid parity in the US: The US National Renewable Energy Laboratory (NREL) has studied the timing of when solar would reach grid parity in different US markets,[26] based on the local retail price of electricity. The results are shown in Figure 8.7, with grid parity (at the retail level) being the dark red color. The map on the left in Figure 8.7 shows the status of grid parity in 2008, when solar TIC was about $8/watt. Note that grid parity occurred generally in states with high electricity costs, such as New York, California, and Hawaii. The map on the right shows the anticipated areas of grid parity in 2014 if the installed price of solar is lowered to $3.50/watt. That is a conservative estimate; it could be $3.00/watt by the end of 2014. For comparison's sake, in 2012, the TIC in Germany was about $2.35 for small (<10kW) residential systems.[27] It's striking how much more of the United States could be at grid parity by the end of 2014.

2008 **2014**

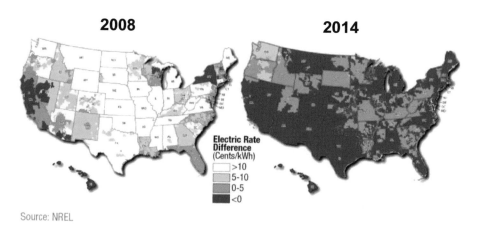

Source: NREL

Figure 8.7 – Grid parity in the United States 2008 vs. projected 2014

To put this in perspective though, the map only indicates the *potential* market for solar. Just because solar reaches grid parity does not mean that it will necessarily be adopted. As NREL points out:

> *Overall, the scenarios evaluated represent a market entry point for solar PV. However, the scenarios do not consider the potential for a deep, sustained market. PV break-even does not imply that customers will necessarily adopt PV, and only a fraction of customers in each utility will have the necessary combination of good solar access and attractive financing options to consider a PV system.*

The topic of how to translate grid parity into demand will be covered in the next chapter.

Global grid parity: Is the US ahead or behind other countries, in terms of grid parity? Yes….both. Countries have different cost structures, and of course the sun shines more in some places than in others, so each country will achieve grid parity at different times. Overall the cost of electricity in the US is relatively low because of abundant coal, gas, and hydro resources. So the fact that the US is approaching wide-scale parity means many other countries are as well.

A 2011 study done by Q-Cells and the Reiner Lemoine Institute[28] estimated that 28 countries would be at grid parity by the end of 2012. In fact, by 2013 an astonishing 102 countries had made it, including Spain, Italy, Germany, Brazil, Denmark, Australia, Chile, Sweden, and Portugal.

A Deutsche Bank report (Fig. 8.8) found that 60% of solar markets depended on government incentives (which they label as "unsustainable") in 2012, but that number was falling fast. By the end of 2014, the report estimates that three-quarters of the markets will be "sustainable," needing little or no subsidy to be economically competitive.

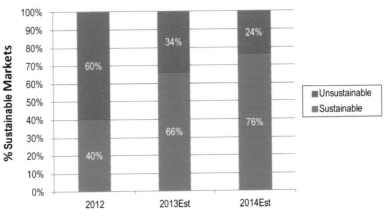

Source: Deutsche Bank

Figure 8.8 - Percent of global markets at grid parity w/o incentives

This is probably the strongest endorsement to date by a third party that the industry has passed the tipping point, and is entering a period of long-term, sustainable growth.

~ ~ ~ ~ ~ ~ ~ ~ ~ ~ ~ ~ ~

This is the beginning of Solar 2.0.

Chapter 9

Solar 2.0
Growth and Turmoil

The passing of the tipping point is driving rapid market growth as we move toward grid parity in more of the world. In this commercialization phase of Solar 2.0, innovations to reduce cost will be the driving force, with growth coming not from policy but from sustainable economics. This shift will be exciting but tumultuous, accompanied by massive consolidation, vertical integration, and cyclical growth spurts. Lessons from the automotive and semiconductor industries show us that these changes are to be expected, as solar becomes an integral part of the world's new energy economy.

The birth of an industry is exciting, and solar is no exception. Nascent industries typically go through the classic "S-curve," starting out slowly and then accelerating until maturity is reached and growth slows to that of the global economy. As we pass the tipping point in solar and grid parity is being realized in more of the world, growth will be driven by innovations that lower cost.

This growth will continue for decades as solar becomes increasingly important to the global energy mix. I rode this curve in the semiconductor industry from the 1970s thru the 2000s. It was exciting, crazy, and world changing. I expect solar's emergence will be the same.

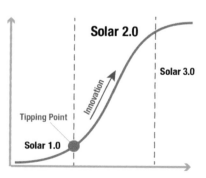

Rapid and Sustained Growth

What can be expected in this next phase of solar? The first thing we expect is a long period of sustained high growth. We can make reasonable growth estimates using two different reference points: a bottoms-up analysis using residential grid parity projections for the US (where the data is most readily available), and a top-down analysis based on projections of how the future energy gap will likely be filled.

Richard Keiser, at the consulting firm Keiser Analytics,[29] looked at the US by area code to determine the local residential cost and potential demand for solar electricity, both with and without the 30% Investment Tax Credit (ITC) offered by the US government (Fig. 9.1).

Installed Cost ($/w)	With ITC (GW)	Without ITC (GW)
$5.50	3	0
$5.00	5	1
$4.50	20	1
$4.00	33	2
$3.50	99	4
$3.00	308	20
$2.50	493	53

Figure 9.1 – Potential US solar demand with and without ITC

This analysis uses US residential installed costs, which at the end of 2013 were about $4.50/watt. Still, what's interesting is that at $4.50/watt (and with the ITC), Keiser estimates solar could potentially support 20GW, in terms of being economically competitive with local conventional sources. But in 2013 the US installed less than 5GW, or 25% of the potential market. Cost, then, is not the problem; other factors, such as finance, marketing, and resistance to distributed generation are limiting deployment.

Now let's look at what could happen if the US reduces total installed costs to German levels of less than $2.50/watt. In that case, the US potential market could be more than 400GW with the ITC, and 50GW without it. Supporting this estimate is a late 2013 Deutsche Bank forecast that the market for solar in the US could exceed 50GW in 2016.[30] This would require over 100% growth per year and may not occur, but again...clearly, the limitation on solar adoption won't be cost.

From a global viewpoint, a top-down analysis can be done based on the demand for renewable energy overall. There is an emerging energy gap of 46,000GW (see Chapter 2). Most analysts agree that renewables cannot fill that entire gap; forecasts range from renewables accounting for 10% of the total mix to well over 50%. Here is my take (Fig. 9.2), based on multiple sources that take into account the needs of the market, the limitation of the technologies involved, the relative costs, and a roughly middle-of-the-road growth projection.

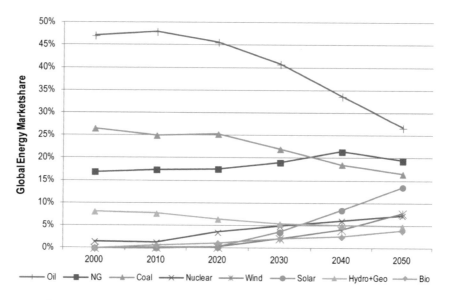

Figure 9.2 – Global energy mix, 2000 to 2050 (author's forecast)

In this scenario oil and coal decline relatively slowly, as natural gas and other renewables take up the slack. Solar is the fastest growing renewable, and accounts for 14% of the energy mix by 2050. If that sounds disappointingly low, consider that reaching that level requires an additional 15,000GW of installed capacity. To put that in perspective, in 2013 the solar industry celebrated reaching 100GW of cumulative installed capacity. So there's a mere 14,900GW to go. Is that doable? Yes: getting there will only require a compounded annual growth rate of about 10% per year. Considering the industry has been growing about 50% per year for the last decade, it doesn't seem out of the question.

The following chart gives my projections for growth over the next decade.

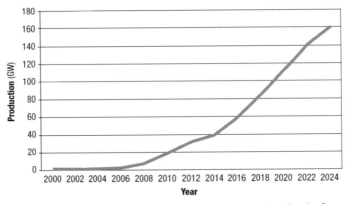

Figure 9.3 – Forecast of global solar installations (author's forecast)

Keep in mind, that while my estimates might be considered aggressive, almost every solar forecast for the last 10 years has *under-estimated* growth. In my opinion, solar is in for a wild ride for the next 30-40 years, and there will be pain as well as gain.

Consolidation

What kind of pain? For a start, the industry is going to consolidate. Henry Ford famously said that if he had listened to consumers, he would have built a faster horse. Ford helped to build a global industry that was based on providing a universal need — transportation — fundamentally differently. Solar, too, wants to supply something basic — energy — in a better way. When growth accelerated in the auto industry in the early 20th century, the number of companies fell rapidly (Fig. 9.4).

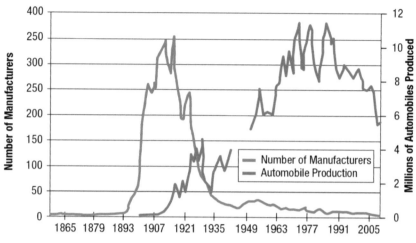

Figure 9.4 – Number of car companies vs. car sales volume

Why did consolidation and growth go together? You could tell the story of the auto industry this way:

> *A new technology surfaces. It is exciting and attracts a lot of investment, so many companies spring up to take advantage of the opportunities. But cars are expensive, so sales are limited. All the new companies experiment with different ideas about how to meet the needs of the market. Then Henry Ford breaks the code with his low-cost assembly line, and drives costs way down. This makes the car much more affordable and sales rise, but higher-cost competitors cannot keep up. They are acquired or go out of business.*

The same thing is happening in solar. From 2005 to 2010, at the height of cleantech popularity, venture capital invested more than $6 billion in the solar industry. Hundreds of new solar companies got started. Then China came in, disrupting the solar industry with the same profound consequences that Henry Ford brought to autos. By adding an enormous amount of capacity (30GW), China drove prices down and over a hundred companies went out of business.

The collapse was epic.

The first major failure was Solyndra in 2011, representing more than $1 billion of investment. Then as the manufacturing base moved from Germany to China, a number of European companies went out of business. The #1 solar company in the world in 2008 (Q-cells) failed, and was eventually bought out by Hanwha of South Korea. In the US, VCs couldn't find exits, either through IPOs or M&A (mergers and acquisitions), and therefore could get no return on their investments. I remember one limited partner investor telling a VC, "If you invest in any more solar deals, you'll never see a dollar from me again!" By 2012 new solar investment was drying up, and the culling continued, albeit at a slower pace the following year.[31] Even China saw hundreds of failures, including the company once considered the country's crown jewel, Suntech. After reaching #1 market share in 2011, Suntech shocked the world by going bankrupt in 2013. By 2014, almost 80% of the companies that were around at the peak in 2008 were gone (Fig. 9.5).

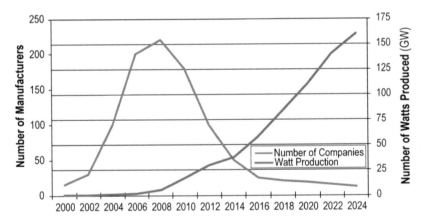

Figure 9.5 – Number of solar companies vs. solar volume (internal data)
Note: Number of companies represents only thin film companies as the status of silicon companies in China is often difficult to reliably verify.

The consolidation is not over. In fact, it has just started. Most of the failures have been startups or small companies with little capacity. If consolidation is measured by percentage of market share of the top ten companies, then we have actually gone in the opposite direction. The top ten solar suppliers have reduced market share over the last few years from 55% in 2010 to less than 41% in 2012. Since the small guys are definitely falling off, if the Tier One (top ten) suppliers are not gaining share, it means the mid-size players must be picking up the slack. Consolidation, therefore, is not done yet. Just as in Henry Ford's time, only those companies that can succeed at the lowest costs will make it. Somewhere around 10 to 20 players will likely control most of the global market.

Because of consolidation, the solar survivors are bigger, on the whole, and they need to get bigger still (Fig. 9.6). Most solar companies today produce in the megawatt range; only the industry leaders have gigawatt capacity. In Solar 2.0, the smallest producers will be in the gigawatt range, and the leaders could produce 10, 20 or even 30GW. This capacity requirement will drive not only factory design, but factory location as well. Since it is risky to locate so much capacity in one location or on one continent, the successful companies in the gigawatt era of Solar 2.0 will become truly multinational.

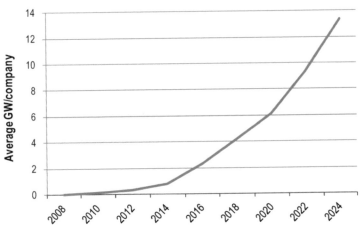

Figure 9.6 – Average volume of solar companies (internal forecast)

There is an important lesson to keep in mind here. The growth cycle in the automotive industry did not begin because Henry Ford made the fastest car, or the prettiest one. It was the *cheapest*. When Ford first introduced the Model T in 1908, it was the first car the masses could at least aspire to own, but it still cost $950 (more than the average US salary). That year Ford produced 10,000 Model Ts. By 1924, a Model T cost $300, and Ford was making 2 million of them. It was all about cost, and today that is true for solar as well.

Vertical Integration

Solar companies are also becoming more vertically integrated as they acquire companies up and down the supply chain. Similar to what Ford did in the early 20th century, when it ended up owning everything from iron mines to car dealerships, solar companies are buying suppliers on one side and their customers (solar project developers) on the other.

Figure 9.7 shows how various solar companies became more vertically integrated from 2004 to 2012. The upstream supply segments are shown on the left, with the downstream, customer-facing segments on the right.

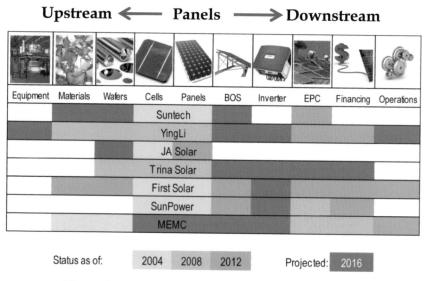

Figure 9.7 – Vertical integration, progress over time

In 2004 (lightest green) the industry was relatively simple, with companies performing only a few functions, mostly cell and panel manufacturing. In 2008 (darker green) upstream integration began. For example, YingLi entered the market with both wafer and polysilicon production. At the same time, downstream integration was getting started with both SunPower and MEMC adding EPC (engineering-procurement-construction) capability. By 2012 this downstream trend was in full swing.

First Solar pulled off what was probably the most aggressive vertical expansion. Fueled by a high stock price, it acquired companies all along the value chain—everything from tellurium mines to solar development projects to inverters and BOS companies. In doing so, First Solar essentially "bought their customer." Acquiring a project development company with a backlog of solar power plants, they were able to assure a pipeline for their CdTe panels. And when cheap Chinese silicon panels flooded the market in 2012, First Solar was immune because they had locked in contracts to supply their own projects with panels.

First Solar is now considered a solar developer, and is one of the world's most profitable solar companies. Their success has inspired Tier One Chinese suppliers to copy them, by aggressively integrating downstream. In Solar 2.0, this trend will continue with companies vertically integrating up and down the supply chain. By 2016, my forecast is that at least one company will be fully integrated across every segment.

There are a variety of reasons to pursue vertical integration: capturing a strategic supplier, assuring a sales pipeline, optimizing system design, or just absorbing as much margin as possible in the value chain. In the past, vertical integration was mostly a matter of reducing the costs of panels, and was thus focused on the upstream segment. That meant acquiring both silicon wafer companies and polysilicon feedstock suppliers.

More recently though, there has been activity on the downstream side, in an effort to buy profits. NREL has estimated where profits exist in the solar value chain (Fig. 9.8).[32] One notable insight from this chart is that panels are amongst the least profitable areas.

Therefore, panel manufacturers are motivated to acquire upstream or downstream to maximize profits.

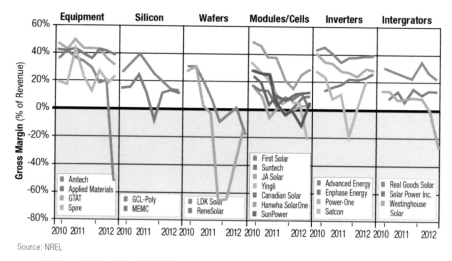

Figure 9.8 – Profitability (gross margin) in the solar supply chain

A Cyclical Industry

As mentioned before, this growth will not be without turmoil. It is likely to have some difficult periods. Figure 9.9 shows the cyclical history of the semiconductor industry; solar looks to be replicating this painful cycle almost to perfection.

Source: Brightside Analytics (DRAM), Brad Mattson (Solar)

Figure 9.9 – Semiconductor memory cycle

The beginning of the cycle is when an industry is labeled "strategic," and government policy encourages investment. This often leads to over-enthusiasm and a fast-growing market (the "Robust Market" arrow). Such growth drives capacity additions (Capital Equipment Expansion), as everyone wants a piece of this exciting new industry. New entrants come in and existing companies expand in order to maintain or increase market share (Excess Capacity). This drives prices down (Price Softening) as companies try to reduce inventory. But then companies cannot maintain profitability, and cash-flow problems slow down or halt the expansion (Weak Market). Manufacturers stop buying equipment, and some go out of business (Capital Equipment Contraction). As a result, supply slows down, stops, or even reverses as capacity comes off-line (Little Capacity Added).

In parallel with these dynamics on the supply side, the Price Softening phase of the cycle also affects demand: Lower prices increase demand. This was true in semiconductors, and is certainly true for solar. As demand increases, we begin to see prices firm up (Price Firming and Stability), profits come back, and the market looks good again. Then, inevitably, companies decide to increase capacity to gain market share. Ta-da! We are back at the Robust Market arrow.

China initiated this cycle with its massive overinvestment in capacity; this led to the huge collapse in 2010-12. Not only was the sun not shining on solar any more, but some people were sure the sky was falling. It all sounded very much like the wails of the 1980s, when people worried that Japan was taking over semiconductors. Just as semiconductors evolved to be a truly global industry, so will solar. China has grabbed the manufacturing lead, but the game is far from over, and there is plenty of room for global competition.

In 2014, the market is still a little weak on the supply side, but almost all the overcapacity is gone. As predicted by the cycle, the price softening phase was followed by increased demand-driven growth of over 50% in many markets. Prices firmed up in Q1 2013, and have been stable since. As of early 2014, we are through the worst part of the cycle. No question: These past few years have been difficult. But they are also characteristic of an emerging industry, part of a healthy dynamic that drives costs down, and thus will enable solar to grow.

~ ~ ~ ~ ~ ~ ~ ~ ~ ~ ~ ~

Solar 1.0 was driven by technology and policy, and innovation was mostly about the hardware. Solar 2.0 has to be about costs, competitiveness, and innovation that touch the entire ecosystem.

In the next section of this book, I outline the opportunities for innovation and cost reduction—the fundamentals that will enable the industry to fulfill its potential in Solar 2.0.

Chapter 10

Introduction to the Roadmaps

Now we transition from what we know about solar to what we can do about it. Solar has accumulated a fair amount of capital: people, money, university research, supportive politicians, and a great degree of public support. But these resources have been spread so thinly that the impact has been limited. It makes sense to aggregate these resources, and then channel them to the most promising avenues for success. Think of the industry as a redwood tree: One must nurture a young seedling with abundant water and fertilizer so that it grows into a towering redwood. But suppose we take the nurturing water and fertilizer, and spread it over an entire forest instead. We may not get a towering redwood; more likely we'll end up with a field of daisies or, even worse, weeds.

Will we spread resources thinly, or concentrate them where they'll have the most positive impact? We can't afford to sink any more private or public money into dead-end strategies. Tough choices are required. The following chapters represent those tough choices in a series of roadmaps. These eight roadmaps transform the vision of a bright solar future into a step-by-step "how to get it done" guide. This is where the rubber meets the solar road. The roadmaps are:

Technology (materials and devices): This roadmap forecasts the successful solar materials technologies and recommends that only three be pursued: c-Si, CdTe, and CIGS. Winning cell and panel architectures are also discussed. If consensus can be achieved in these areas of technology, it is more likely that R&D investments will be channeled to achieve the most successful results.

Efficiency: This provides an efficiency forecast for each of the technologies and suggests "best practices" to achieve forecasted efficiencies, including the development of unit and cost metrics.

Manufacturing/cost reduction: Energy is a commodity business; therefore, cost is everything. This roadmap suggests where future cost reductions will come from in solar panel manufacturing. Materials, labor and overhead costs will be reviewed in detail.

Products: While the manufacturing roadmap deals with panel costs, the product roadmap deals with the entire solar power system, and suggests how to deal with BOS and other soft costs. It also addresses how systems can evolve to fit specific markets and applications.

Business models: Solar 2.0 will be as much about market innovation as technical innovation. This roadmap explores those possibilities, such as business models that address distribution, finance, customer acquisition, and other costs. The difference between c-Si and thin film business models is also explored.

Investment: This roadmap outlines how, when, and where solar investment is likely to be profitable; how risk mitigation might be improved, and explains ten sensible investment themes.

Policy: What kind of support is needed for further research and deployment? What policies will help to create jobs and lead to energy independence? These and related questions are the focus of the policy roadmap.

Public support: The audience for this roadmap is, essentially, everyone, because only with public support can the changes spelled out in the policy and investment roadmaps be put into effect. It is up to all of us to make it happen.

Note: These roadmaps are sometimes very detailed, written for a specific audience: technologist, policy-maker, educator, manufacturer, solar customer, or just a concerned citizen. If an area is not of interest, feel free to skip around.

~ ~ ~ ~ ~ ~ ~ ~ ~ ~ ~ ~

Obviously, these topics are complicated, and these chapters might not do justice to the depth of discussion required. The intent is to provide a "strawman" to initiate that discussion; to help us all move toward a consensus on how we might jointly proceed. I don't presume to have all the answers, but hope to start a conversation about how to concentrate resources to create the best chance of success.

Let us know what you think at our website: thesolarphoenix.com.

Chapter 11
Technology Roadmap

This chapter explains and then forecasts the most promising technologies for solar panel materials and architectures. The industry cannot afford to pursue all possible technologies; we need to make choices. On the material side, the winners will be c-Si, CdTe and CIGS. On the architectural side, "monolithically integrated" panels will eventually prevail over "singulated cells." Although c-Si won Solar 1.0 and will remain strong for the foreseeable future, thin films — specifically CdTe and CIGS implemented with monolithic integration on glass substrates — hold the most long-term promise.

How will technology change in the era of Solar 2.0? The biggest shift will be that of focus, changing from technology that expands the frontiers of science, to technology that is economically manufacturable. Cost is everything in Solar 2.0, and the technologies that reduce cost will succeed. This chapter is less of a "how-to" guide for those technologies, and more of a roadmap of which are most likely to succeed. The idea is to direct R&D resources to the areas of highest potential.

There are two broad categories of technology in solar: materials and architecture. "Materials" refers to what the solar cell is made up of. Examples are c-Si, CdTe, CIGS, GaAs, CZTS, organic, and dye-sensitized. "Architecture" refers to the design of the solar cell or solar panel. Examples include dual and triple junction, nano-structures, rectennas for cell architectures and singulated cells, monolithic integration, flexible, and BIPV for panel architectures. The materials and architecture interact and together determine the performance of a solar cell, so while this chapter looks at each separately, there will be overlap in the discussion.

Materials Technologies

When it comes to technology, sometimes it doesn't matter what is best. Once a particular technology reaches critical mass, it can prevail even if it isn't superior. Sony's Betamax technology, for example, was far superior to VHS. But VHS got to scale faster, and became the industry standard. The first to volume won.

This analogy is instructive when it comes to the solar industry (although the story is a little more complicated). Silicon technology, the original solar material, was the de facto standard, but was not the first to get to scale. First Solar got to scale first (1GW) using a thin film technology, CdTe. Round one to First Solar. But when China entered the fray, it went with silicon because the material and equipment was readily available, whereas First Solar's technology was proprietary. Round two went to China, which hit volumes of up to 30GW.

What about all of the other technologies beside CdTe and c-Si?

- GaAs
- CIGS and CZTS
- Dual and triple junction cells
- Organic PV and dye-sensitized PV
- Nano-structures
- Quantum dots and rectennas

These are interesting, but most of them never got up to scale in Solar 1.0, and it is unlikely they have any chance to catch up in Solar 2.0. The train has left the station.

There is one exception, though. In addition to c-Si and CdTe, CIGS (another kind of thin film) also got to the gigawatt scale. Combined, these three technologies have received billions in investment and are the likely winners in Solar 2.0. They are therefore the only technologies discussed here. Each has its own strengths and challenges.

Silicon

Silicon was the clear winner at the end of Solar 1.0. It became ever cheaper, leading to high volumes and market dominance. In fact, silicon's market share is about 90%. But what's next? C-Si is a mature technology and is just about fully optimized. The future, then, may lie in new cell architectures. The NREL cell-structure in Figure 11.1 illustrates some of the advanced silicon technologies that may be implemented over the next five to 10 years. This cell is similar to the Passivated Emitter and Rear Cell (PERC) design, which includes a reflector on the back metallization to increase light absorption, and has a selective emitter. This NREL solar cell is expected to achieve an efficiency of 21%, and yield a panel efficiency of 18.7% That is several percentage points better than the common technology used in today's high-volume c-Si production, and such a cell can be economically produced.

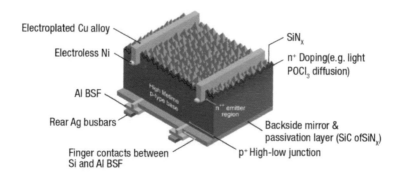

Figure 11.1 – NREL advanced cell architecture

There are many other new technologies being evaluated for making silicon wafers and solar cells. Here are the major ones:

- Fluidized bed reactors (FBR) are a way to make polysilicon. This could cost less than current techniques, and there is movement to this technology.

- Diamond wafer saws: This technology reduces "kerf loss" (silicon lost in slicing of the wafers) and is likely to be used in making sc-Si wafers. Other, lower-cost methods may be used for mc-Si wafers.

- Kerf-less, thin silicon: The idea is to cut costs by reducing the amount of silicon required to create a solar cell. There are high- and low-efficiency kerfless techniques. On the low end, silicon wafers are cast or deposited on low-cost substrates. On the high end, high-efficiency cells are formed on the surface of sc-Si wafers and then a thin film of silicon, including the solar cell, is separated from the original wafer. The separation process could be through a lift-off or cleaving process. This high-end method is typically expensive.

- Monocast: This is a new method for casting the multicrystalline ingots to achieve properties closer to monocrystalline silicon. Because mono has higher efficiency, the hope is to increase efficiency without the huge increase in cost associated with mono-crystalline ingots.

- Selective emitter: This technology is used to reduce contact resistance with silver grids while maintaining higher resistivity material in the bulk to reduce recombination losses. The benefit is improved efficiency (up to 0.7%),[33] which is a relatively large gain. As a result, selective emitter technology is being implemented in many production lines.

- Double print: The silver grid lines used to collect the current block sunlight and reduce efficiency. By making the lines taller and narrower, the same current can be collected with fewer "shadowing" losses. Double printing, where the line is created in two steps, forms a taller, narrower line. There are other techniques to reduce shadow losses by improving the conductivity of the lines so they can be narrower.

- HIT technology: HIT stands for Hetero-junction with Intrinsic Thin layer. This design, like the PERC cell, achieves higher efficiency, but does so with a more complex process flow. The subsequent increase in production costs makes it noncompetitive with multi-crystalline silicon; therefore it is only used in niche, high-efficiency, limited-area markets like small roofs or consumer electronics.

- Emitter wrap-through: This maximizes efficiency by moving the contacts to the back side of the wafer so there are no shadow losses. This technique, used by SunPower, is expensive and applicable only in niche markets.

A more detailed discussion of silicon technology could take up its own book, so it will not be attempted here. (Those who want a more detailed Silicon Technology Roadmap should look at DOE's EERE group[34] and SEMI's International Technology Roadmap for PV).[35]

Thin Film

Considering c-Si's dominance, it's fair to ask: "If silicon won Solar 1.0, doesn't that mean that thin film lost?" The answer is, not quite. While many thin film companies have failed, First Solar has been the most consistently profitable solar company to date, and for many years had the industry's lowest production costs.

Theoretically, thin film should be cheaper than silicon, which is why it garnered so much support in the past, and why many people still think it is the future. After all, it requires 100 times less material to absorb the same amount of sunlight as c-Si, and it can be deposited on glass, which is much cheaper than silicon. In spite of its inherent cost advantage, the massive investment in c-Si has driven many thin film companies out of business, and there will be further consolidation. Only a few thin film companies will make it — perhaps no more than five. Let's look at the two most promising thin film technologies: CdTe and CIGS.

CdTe: CdTe (cadmium telluride) was the first PV technology to reach scale, and it will continue to be a force. The problem with CdTe is it that it has the lowest efficiency of the three contenders. But it is also the cheapest to make, and performs better than c-Si in terms of solar yield. CdTe degrades less than any other technology, as shown in Figure 11.2.

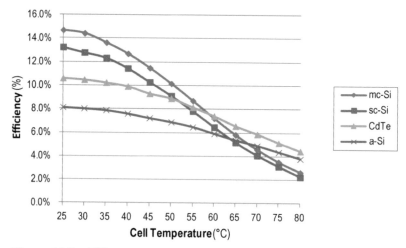

Figure 11.2 – Efficiency degradation vs. operating temperature

At room temperature (25°C) CdTe has almost the worst performance, but at elevated temperatures (above 60°C) the efficiency is the highest. Since panels tend to be installed in areas of bright sunlight they do tend to get hot, so this is not a trivial advantage. In fact, enhancing its weakest point—efficiency—could be the salvation for CdTe.

The CdTe champion, First Solar, is one of the industry's strongest companies, with an enviable balance sheet. In 2013 it had revenues of $3.3 billion, and was sitting on more than $1.5 billion in cash.[36] First Solar has scale and a high degree of vertical integration. It does so many things well that it is going to be difficult for a competitor to muscle into the CdTe space.

There is one possibility. A possible weakness of First Solar is the small size of its panels—only 0.7m², which is less than half the norm (1.6m²). This is a limitation of First Solar's manufacturing process, not a fundamental limitation of CdTe. As with semiconductors, larger panels in solar will improve productivity by far more than they increase costs. A CdTe company that beats First Solar to 2.0m² panels could give it a run for its money.

CIGS (*Copper-Indium-Gallium-diSelenide*): CIGS will not only survive in Solar 2.0, but it will likely become the leader in thin film technology (and possibly, all of solar). This is a bold prediction, but there is a sound basis for it. In thin film, CIGS has the highest-efficiency panels on the production floor, as well as the highest efficiency cells in the R&D labs (see next chapter). A less obvious advantage of CIGS is cost. Although today it is the most expensive of the three finalists, CIGS has the ability to perform high-throughput processing with the lowest materials cost. It just has not yet reached the scale of c-Si or CdTe.

CIGS does have its negatives. The technology is not as well understood as others, and the efficiencies achieved in the lab have not yet made it into high-volume manufacturing. There is also the matter of complexity. There are many different methods for depositing CIGS, and there are a variety of substrate materials and device architectures. There are numerous front and back contact techniques with different structures, resulting in different manufacturing process flows.

This flexibility is both a blessing and a curse. There is enough variety that a low-cost manufacturing process will eventually be found. But the curse is that there are also many ways to fail—and these failures (Nanosolar, SoloPower, Solyndra, etc.) have had a lot of attention. On the "blessing" side are the processing options. The deposition method known as co-evaporation is where the four CIGS materials—Copper, Indium, Gallium, and Selenium—are deposited by thermal evaporation at the same time (hence the name "*co-*"evaporation). Co-evap offers one of those unique opportunities in physics: a process that is the fastest, highest quality and lowest cost, all at the same time. CIGS implemented with co-evaporation yields both the highest-efficiency devices and the fastest manufacturing process. Moreover, any evaporative process is also a distillation process, and therefore self-purifying, meaning that it is possible to use less refined (less expensive) source materials. The fastest processing with the cheapest materials should translate into the lowest-cost manufacturing, but this has yet to be proven.

Another "flexibility" advantage of CIGS is that it can adapt to roll-to-roll (R2R) processing. Let's use printing newspapers as an example. If they are printed a page at a time, efficiency comes by trying to do that really fast. Alternatively, one could print the pages continuously on a high-speed R2R printer—a process that is much faster and cheaper. Commodity products in many industries, such as paper, textiles, and steel, scale up by converting to a R2R process.

Neither CdTe nor c-Si can effectively use R2R processing. Silicon is thick and inflexible, so it can't be rolled. Technically, CdTe could be deposited on a flexible R2R substrate. But this doesn't work well, because the only successful CdTe process to date uses glass substrates that are inflexible and cannot be rolled. R2R processing is possible with CIGS, and has been demonstrated on thin, flexible aluminum, stainless steel and plastic foils. This matters because R2R processing could lead to lower manufacturing costs.

In strictly technical terms then, co-evaporated CIGS looks like the outright winner. The problem is, in spite of billions in investment, no one has made it work yet. Just when it seemed multiple startup CIGS companies might have broken through, the Chinese bulldozer came and squashed most of them. A lot has been learned from those billions invested, but a second generation of tools, processes, and techniques will be needed to revitalize this exciting technology. When the dust settles, it is likely that three CIGS companies will succeed in Solar 2.0—one incumbent and two newcomers (for detailed profiles of the thin film companies that might prosper in Solar 2.0, see Chapter 16).

Of the incumbents, Hanergy is likeliest to survive, simply because it has amassed so much technology and capability, in part through its acquisitions of MiaSolé, Solibro, and Global Solar. Hanergy has a charismatic and dedicated chairman, solid manufacturing capability, and a strong intellectual property (IP) portfolio. Among the technologies Hanergy owns are co-evaporated and sputtered CIGS, stainless steel and glass processing equipment, rigid and flexible substrates, and both MLI and singulate process flows (explained below). Finally, and maybe most importantly, Hanergy has a lot of money.

What about Japan's Solar Frontier, the world leader in CIGS? It has high efficiency, the largest manufacturing capacity, and a rich corporate parent, Showa-Shell, which invested $1 billion to scale up manufacturing to almost a gigawatt in capacity. In post-Fukushima Japan, there has been a renaissance of solar, and with the availability of high domestic subsidies and a "buy Japanese" mentality at home, Solar Frontier is in the right place at the right time. Despite all of this, its prospects are murky. The problem is that Solar Frontier has a very high cost structure—maybe the highest of any thin film company. That is an undesirable trait in a commodity market.

In the long term, it is not clear if Solar Frontier will ever be cost-competitive. It is difficult to get detailed information, but its factory cost seems to be quite high, at approximately $1/watt. That is about 30 percent higher than Nanosolar, and twice as high as MiaSolé, which both failed. Solar Frontier also uses very expensive deposition techniques, and its device architecture requires three sheets of glass (others use one or two). Once the subsidies in Japan run out, it's not clear that Solar Frontier can compete.

On the materials side, then, the roadmap is simple. The likely technologies are c-Si, CdTe, and CIGS. Of these, c-Si is dominant for now. But thin film could catch up and even take the lead in solar 2.0, so there will likely be some successful newcomers in this space. Both thin film technologies, CdTe and CIGS, have a good shot. But CIGS, while less mature now, may have the brightest future of all.

Solar Cell Architecture/Solar Panel Architecture

In terms of both cost and performance, architecture (the structure of the solar cell or solar panel) matters more than materials selection. There are several architectural choices for both cells and panels.

Solar cell architecture: A solar cell is basically a diode — an electrical device formed by joining two dissimilar materials to create a semiconductor "junction." It is in this diode that light is converted into electricity. The architectural choices involve how to form the junction, and how to make electrical contact to the diode so the electricity created can be used.

Of the junction options, there are single-junction devices as well as the more complicated dual- and triple-junction devices. Here we'll discuss only single-junction devices as the more complicated and expensive structures have already been left behind in the race to scale.

In terms of making electrical contact, the front-contact design has the most bearing on cost and performance for both silicon and thin film solar cells. Unfortunately, the lowest resistance (best) contact materials are not transparent, and block the sunlight trying to get into the solar cell (shadowing). The goal is to make narrow lines or transparent contacts to maximize light pass-through, while minimizing contact resistances. Many of the more exotic silicon cell architectures are designed to eliminate the front contact altogether, moving it to the backside to reduce those "shadow" losses, and thus improve efficiency.

Attempts to reduce losses include improving the traditional silver contact grids, improving the front transparent conductive oxide (TCO), applying decals with a wire grid structure, and metal wrap-through (MWT) technology. Results have been mixed. One thing, however, is now clear: MWT is too expensive. Nanosolar went to its corporate death trying to make it work. Of the other options, advanced TCOs for thin films and wired grid structures for c-Si seem most promising.

Solar panel architecture: This may be the most important technical decision a solar company makes. There are many choices such as frame vs. frameless, and panel size and shape. But the most important architecture choice is between singulated cells (SC) or monolithic integration (MLI). The SC architecture involves making relatively small solar cells, and then assembling them into panels. There may be between 50-100 solar cells in one panel. MLI, on the other hand, starts with the full-sized panels and monolithically builds the solar cells on the panel. With MLI there are never individual cells.

MLI is a simpler, cheaper process because there are fewer manufacturing process steps. At first glance, then, MLI looks better than SC. But of course, it's not quite that simple. For one thing, the SC approach is the only way to make c-Si solar panels. SC is now at huge scale, and costs have come way down. Also, SC is compatible with R2R processing, and MLI is less so.

Since the whole panel is processed at the same time with MLI, good process control is required over a much larger area, and an entire panel may be scrapped if one section is bad. SC is more forgiving — weak cells can simply be scrapped and replaced by good ones. So while MLI does cost less to produce, the final cost may be higher due to lower yield.

The bottom line: Since MLI is cheaper, it is probably the better long-term choice — if the yield catches up to SC manufacturing levels. This should occur as processes mature and there is improved quality control. In the thin film market, the verdict is clear: The top two thin film companies and the only ones at scale (First Solar and Solar Frontier) both use MLI.

Of the other panel architectural choices there is a move to larger panels, as this reduces cost. The industry standard is moving toward 1.6m², but there are already 2.0m² panels on the market. Another trend seems to be going from framed to frameless designs. The frames add cost and often reduce reliability, so frameless designs will likely prevail.

~ ~ ~ ~ ~ ~ ~ ~ ~ ~ ~

Silicon technology (c-Si) is the obvious winner so far. Thin film (CdTe and CIGS) will survive, even though many thin film companies will not. The cycle of failure and consolidation will continue, as is always the case as a technology evolves. The result will be fewer but stronger companies.

Silicon will continue to evolve, but it is a very mature technology. There will be incremental advances that squeeze out a few more cents in cost reductions, but nothing-game changing seems to be on the horizon.

Thin film is tougher to call. First Solar is in great shape, and will continue to build on its successful business model. There may be room for one more CdTe company, if someone figures out how to capitalize on First Solar's limitations.

Finally, it is possible, even likely, that two or three CIGS companies will make it. If one of them figures out how to get high yield and high efficiency with MLI panel architecture, this company could even become a leader in Solar 2.0. Thin film, with its inherently lower cost structure, may be the solar technology of the future.

Chapter 12
Efficiency Roadmap

Efficiency is important, indeed critical. There are several types of efficiency including maximum theoretical, world record, and average production, to name a few. Each of these is analyzed and forecasted. The efficiency forecast shows how the thin film materials, CdTe and CIGS, are catching up to mc-Si efficiency, and may pass it in the future. As a final note, although efficiency is important, cost is even more important. Too often, the industry has over-emphasized efficiency, and underemphasized economics. Efficiency improvements need to be compared to the manufacturing costs needed to obtain them.

Efficiency is a ratio; it expresses the amount of energy created by a solar panel relative to the total amount of energy in the sunlight that hits it (defined as 1000 watts per square meter). Think of it as energy out versus energy in. In effect, just shine light on a solar panel and measure the power output. So if a one-square-meter panel produced 1000 watts, that would be 100% efficiency. The norm is more like 100-200 watts, or 10% to 20% efficiency.

That's a simple explanation, but of course things are not that simple. Efficiency can be measured on any sample size, from a 0.5cm² R&D cell all the way up to a 2m² panel, a 40,000X increase in size. Even for the exact same technology, the efficiency will not be the same over this huge range. There is also something known as "aperture" efficiency, in which non-producing areas of a cell or panel are masked so that only the "active areas" are measured. This results in a higher (but unrealistic) efficiency figure. Efficiency can also be measured under different lighting conditions and over different periods of time to derive average values.

So when we talk about efficiency, precision matters, not only to understand the concept, but also to ensure that we compare apples-to-apples. This chapter focuses on three types of efficiency: maximum theoretical, world record, and average efficiency in high-volume production.

Maximum theoretical efficiency: This efficiency is directly related to a physical property of the material called its "band gap," which is measured in electron volts (eV). This property indicates how well and what type of light a material absorbs. The maximum theoretical efficiency for most PV materials, as shown in Fig. 12.1, is between 25% and 30%. For example, consider CdTe. It has a band gap of 1.45eV, which intersects the curve at about 30%. That is its maximum theoretical efficiency.

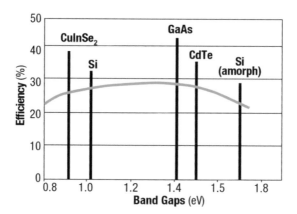

Figure 12.1 – Maximum efficiency vs. band gap

The maximum efficiency is limited because each material only absorbs specific wavelengths of light, and even much of that is converted into heat. Clever work is being done with multiple and graded band gaps to capture more wavelengths and the lost heat energy, raising the maximum for multi-material systems to almost

50%. Unfortunately, these advanced techniques are so expensive they have not seen widespread use in commercial applications.

World record efficiency: Think of these numbers as the reigning champs – the world record-holders in terms of actual output for each type of material. Figure 12.2 shows how world record efficiency has improved, slowly but steadily, for each material system. The highest efficiencies are for the multi-junction devices. For single-junction devices, silicon solar cells have reached efficiencies of more than 25%; thin film cells range from 10% to 21%. The best single-junction material has been GaAs (28.3%), close to the theoretical limit.

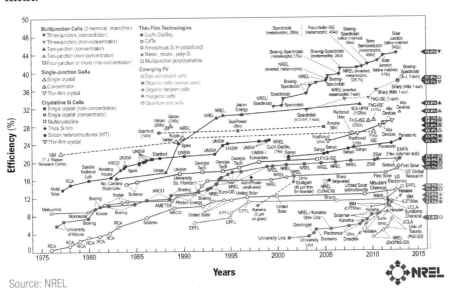

Source: NREL

Figure 12.2 – World record efficiencies for each material system

It's important to note that most of these results were achieved on 0.5 to 1.0 cm^2 solar cells, smaller than a postage stamp. And the results may not be repeatable or scalable. So these world record results often bear little resemblance to what is seen on a commercial production line. Still, these records demonstrate potential and attract attention. If the record is 20% and a company's volume production is 10%, they have reason to believe that there is room for improvement. This concept of "room for improvement" is called headroom, and is a desirable characteristic.

Average efficiency in high-volume manufacturing (HVM): This is the third and most relevant metric. This represents the output of a production line—what an end user gets. HVM efficiency is always less than the record level. The more mature the technology, the smaller the gap is between record and HVM (Fig. 12.3). For example, the gap for the most mature technology, multi-crystalline silicon, is only 5 percentage points; the gap for thin film technologies is about 6 points.

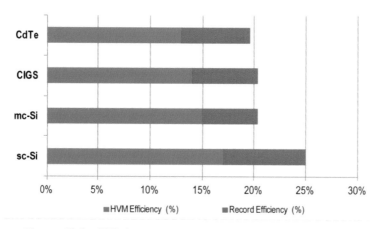

Figure 12.3 – Efficiency gap, world record to average HVM

Efficiency Forecast

World record efficiency is the easiest to forecast because there are 30 years of NREL data from which to draw. These show an improvement of about 1 percentage point every three to four years (Fig. 12.4). The table also shows that improvement has slowed down for most technologies over the last 10 years. This is normal: There is always a steeper slope at the beginning of the any efficiency ramp, and then a flattening in the latter stages of technology development. This occurs because easy fixes are found early, and subsequent efficiency increases are harder to achieve as one gets closer to the theoretical maximums. For mature technologies, such as sc-Si and mc-Si, efficiency has basically leveled off.

		Efficiency		Difference		Rate % inc/yr	
		1984	2014	30 yr	last 10yr	30 yr	last 10yr
Silicon:	sc-Si	16%	25%	9%	0.0%	0.30%	0.0%
	mc-Si	15%	20%	5%	0.4%	0.20%	0.04%
Thin Film:	CdTe	7%	20%	13%	3.6%	0.43%	0.36%
	CIGS	9%	21%	12%	1.2%	0.40%	0.12%

Figure 12.4 – World record efficiency and rate of change

Thin film technology shows a faster rate of improvement, because it started from lower values and increases were easier to achieve. More significantly, it is clear that thin film isn't anywhere close to done. In fact, in the last two years there has been a renaissance of sorts in thin film, with the efficiency of both CdTe and CIGS going up at a rate higher than the "last 10-year" average, or even the higher 30-year rate.

In this race, CdTe stands out. It has improved almost 1%/year over the last 2 years; a phenomenal result. This has been due to the intensive R&D efforts at both First Solar and GE's Primestar (which has now merged with First Solar). But this should be seen more as catch-up than signs of a breakthrough. Before this burst of R&D activity, CdTe had stagnated at 16.8% for almost a decade. It is likely to revert to the more normal trend line of CIGS. In either case, though, thin film has already caught up or passed mc-Si and is continuing to improve.

Before using this historical data to forecast future record efficiencies, it is crucial to understand the amount of R&D that will be applied to the problem. There are four main sources of R&D funding:

- *Government funding* varies around the world, increasing in some areas and decreasing in others. The best we can hope for is that current funding will be maintained, but it will more likely diminish.

- **University research** has been pretty constant over decades, and no change is anticipated. If government spending declines universities can also be affected, because that is an important source of university funding.

- **Large corporate R&D** is hurting because profit margins for solar have been low. This is especially true for c-Si solar manufacturers, who really need R&D to find the innovation needed to get costs down and profitability up. Thin film companies are spending more on R&D, but there are very few of them.

- **Venture capital** is the hardest-hit area. VC investment in solar has gone from more than $1 billion a year at its peak to next to nothing. This will not change until there are more successful exits like SolarCity. For the next few years, VCs are unlikely to be a major source of R&D investment.

Adding it up, R&D spending is likely to be weak for the foreseeable future. World record efficiencies are therefore unlikely to increase at a significant rate. The forecast is provided in Figure 12.5.

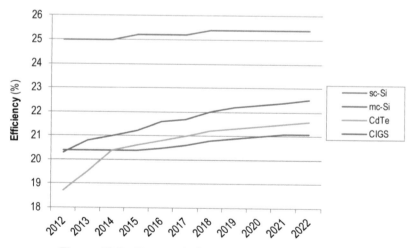

Figure 12.5 – Forecast of world record efficiencies

Progress has been slow for the past 30 years, and there is little reason to expect that to change. The lesson to draw is that believing that efficiency will be the magic bullet to lower solar costs is wishful thinking. Setting new efficiency records might be exciting, but it does not change the slope of the cost curve very much.

And there is another interesting story that shows up in Figure 12.5: thin film's historic crossover. In 2013 CIGS caught up and passed mc-Si in efficiency, and in 2014 CdTe did the same thing (mc-Si accounts for about 60% of the silicon PV industry). One caveat: Both CIGS and CdTe have achieved this only in the R&D lab, not on the production line. Still, this crossover could be huge.

This world record efficiency battle bears watching. Both sc-Si and mc-Si will likely begin to improve again, as silicon companies must respond to the challenge from thin film, and it might be their only way to continue the reduction in $/watt. But they are unlikely to keep up with thin film, which is a newer technology with more room for improvement.

Gap Forecast

To go from world records to forecasting efficiencies on the production floor, the method used in this book is to look at the historic gap between world records and HVM production results, and forecast that gap going forward.

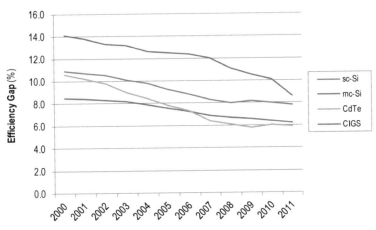

Figure 12.6 – Historic gap, world record to HVM production

Figure 12.6 shows that the gaps have been closing nicely, most dramatically with CIGS, where the gap has closed from 14 percentage points to less than 8 over the last 12 years. As would be expected, the more mature technologies like mc-Si have seen less progress because the gap was relatively small to begin with.

Before forecasting how small this gap will become, it is helpful to understand the bottom limit—that is, the point where physics says progress must stop. These limits exist for three major reasons. First, there is always area "lost" in a panel, due to the need for edge seal, mounting hardware, scribe-lines, and other parts of the panel that for whatever reason do not convert sunlight into electricity. Second, there are losses in interconnecting smaller cells in order to make a large panel. No cell inter-connection scheme, whether singulated cell or monolithic integration, does not suffer some energy loss. Finally, no process can be perfectly scaled—there are always some scaling losses. Factoring all of this in, the general estimate is that the bottom limit of the gap is about 4 percentage points. For example, if the laboratory record is 21%, the best high-volume manufacturing could deliver would be 17%. Using 4% as the absolute limit, the "gap" forecast for each material is shown in Figure 12.7.

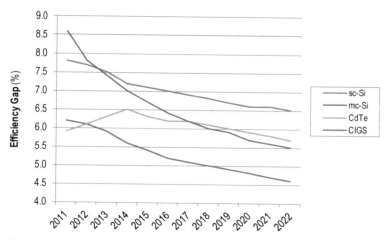

Figure 12.7 – Forecast gap from world record to HVM production

As shown, only mc-Si is likely to approach the 4-point gap limit over the next ten years. There is intense competitive pressure and mc-Si is getting the most time and money. Single crystal (sc-Si) performs worst. To understand why, one must look at the geometry. These wafers start out round, and they do not fit well in square panels. A lot of area is wasted, which increases the efficiency gap. Unless someone invents square sc-Si wafers, it will be difficult for sc-Si to get to a 6-point gap. CdTe and CIGS fall somewhere in between, with 5.5% to 6% losses at the end, but CdTe has a strange trajectory. It is the only technology that has seen the gap increase over the last few years. This is because of the rapid rate of change in the champion results. Since those improvements have not filtered down into manufacturing yet, the gap has been increasing. This will not continue: Manufacturing efficiency will increase, and the rapid champion results will likely slow down. As mentioned before, the recent rise has a lot to do with catching up from an entire decade with little to no improvement. Within a few years, CdTe should regress to the normal downward trend.

HVM Efficiency Forecast

This is the most important forecast since the entire solar industry depends on what comes off the production floor, not what happens in the lab. To develop the HVM forecast, we start with the world record forecast and the gaps (forecasted in Fig 12.7) are subtracted to get the anticipated HVM efficiencies. Figure 12.8 shows the result for each material. Clearly, sc-Si is still well ahead, but thin film technologies are narrowing the difference.

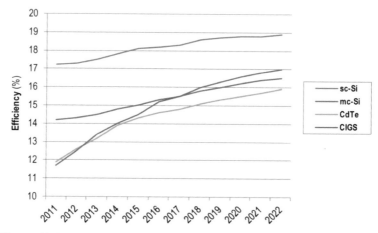

Figure 12.8 – Forecasted average HVM production efficiency

The most stunning event on this graph is the trajectory of CIGS. By 2017, or even sooner, it will be as efficient as the industry volume leader, mc-Si. That is one of the reasons why so many have invested in CIGS. CdTe is doing well. If the recent advancements in world record CdTe efficiency can be translated from R&D to the production floor, it could also pass mc-Si.

How do these forecasts compare to actual data in the field?

Figure 12.9 shows the actual efficiencies for projects installed under the California Solar Initiative (CSI) program over a five-year period. This data serves as a good reality check for efficiency projections.

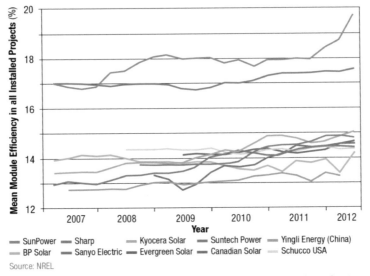

Figure 12.9 – Actual module efficiency on installed projects

There are two efficiency groupings. SunPower leads the high-efficiency group, closely followed by Sanyo, which has a high-efficiency HIT cell. These more expensive panels are used for residential applications with small roofs. Homeowners will pay a premium for SunPower or Sanyo panels to maximize energy production. There are other advanced, high-efficiency devices on the market (PERL, HIT, Pluto, etc.), but because of costs few have made it into high-volume production. Combined, the high-efficiency group has never accounted for more than 30% of total solar production, and that number is going down.

The lower-efficiency data is from a group of mc-Si suppliers. This data, representing the bulk of the market, shows how slowly efficiency is improving. The lower results in the group show only a 0.6% gain over five years, and even the most successful, Sharp and Kyocera, only gained 1.5% over that period. This historical gain matches nicely with the forecasted ~1% gain for mc-Si over the next five years.

A final comment: In spite of the efficiency lead, sc-Si is still losing market share because it has the most expensive manufacturing process, and this is a very cost-competitive market. Its less efficient cousin, mc-Si, is picking up market share every year because it has lower manufacturing costs, and therefore cost/watt. Thin film typically has an even lower manufacturing cost than mc-Si, so if a crossover in production efficiency occurs, thin film might take the lead.

~ ~ ~ ~ ~ ~ ~ ~ ~ ~ ~ ~

Solar 2.0 needs to focus on manufacturing costs, but efficiency is still very important. It is a direct measure of the power output of the system—watts, which account for half of the $/watt calculation. With record efficiencies in silicon improving slowly, there is more to be gained from reducing the gap between champion results and HVM efficiency. That will require a shift in R&D priorities toward manufacturing process stability and process control.

An exception to the "improving slowly" outlook is the forecast for both CdTe and CIGS. These thin films have recently seen rapid increases in champion results, offering the possibility of surpassing mc-Si in efficiency on the manufacturing floor within the next few years. Thin film might be making a comeback.

Chapter 13

Manufacturing/Cost Roadmap

> *To be competitive in solar, low-cost manufacturing is crucial; it is the key determinant of success or failure. In order to reach grid parity, the industry needs to cut costs roughly in half. Each component of costs — materials, labor, and overhead — must be minimized. To operate a factory at world-class cost levels requires a good supply chain, high-volume production tools, a productive workforce, and good management. If these factors are combined with automated (advanced) manufacturing technology, many countries, not just China, can be competitive in manufacturing. With Solar 2.0's automated, giga-scale factories, manufacturing will become more location-independent.*

Cost is everything. That is why the manufacturing roadmap is more important than the technology roadmap. This chapter will focus on the cost of manufacturing solar panels. Reducing the other system costs will be covered in the chapters on the Product and Business Model roadmaps.

To meet the DOE's SunShot goals, panels must be sold for $0.50/watt. Based on 2014 selling prices of more than $0.70/watt, this is doable but difficult. To be sustainable a reasonable profit is required, meaning a manufacturing cost of about $0.40/watt, or almost half of the current level.

Costs: $/watt vs. $/m²

Costs in solar are commonly measured in dollars per watt, but this is really a "performance cost" metric, as opposed to a pure cost metric, because it includes the output of the solar panel (watts). Remember the formula:

$$\frac{\$}{watt} \quad = \quad \frac{\$}{m^2} \quad \div \quad watt/m^2$$

$$\text{(cost)} \qquad \text{(????)} \qquad \text{(efficiency)}$$

What is shocking is that there is no industry-accepted name for the parameter in the middle (dollar per square meter). This is the cost to manufacture one square meter of solar panels. To avoid confusion, the $/m² "pure cost" metric will be called manufacturing cost, and is the focus of this chapter.

Most corporations describe manufacturing costs using the standard term, COGS, or cost of goods sold. Public companies, by law, must include in their COGS calculation all the costs to manufacture a product. They do not have to include sales, R&D, or administrative costs, as these are not part of manufacturing.

Within COGS there are three main categories: materials, labor, and overhead. Materials include all the components (glass, metal, plastic, etc.) that end up in a product. Labor is the direct labor used to make the product. This does not include labor required for materials handling, quality control, maintenance, manufacturing engineering, or management, which counts as part of overhead. Overhead also includes rent, utilities, and depreciation. Depreciation is calculated by dividing the cost of the manufacturing plant (facility and equipment) by its lifetime, usually seven years.

Factory cost is not just important because of its impact on depreciation; it is also of strategic importance because of solar's capital intensity. Capital intensity is a measure of factory cost relative to its output. The metric used is $/watt, with the "$" being the capital expenditure (capex) to build the factory, and the "watt" being the yearly output of the factory. Unfortunately this can be confusing, as this same $/watt unit is used to measure panel cost and installed system costs.

Lower capital intensity is obviously better, as it makes growth, in the form of capacity expansion, easier. For example, if the goal is to add 30GW of solar capacity, and the typical capex of a c-Si facility is $1/watt, then $30 billion is required just to build the factories. At $0.50/watt capex, only $15 billion is needed.

To minimize COGS, the first step is to identify the top cost drivers, then prioritize and work on areas that will make the biggest difference. The costs for silicon (Part 1) and thin film technologies (Part 2) are different, and will be reviewed and forecasted separately. After each section, the efficiency forecast from Chapter 12 will be brought in to translate the $/m² manufacturing cost into the industry darling metric, $/watt.

Part 1
COGS for Silicon: Materials

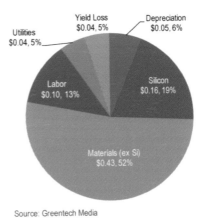

Figure 13.1 is a COGS estimate for a leading Chinese c-Si module manufacturer. The panel cost is $0.82/watt. The red and green sections make up the materials, and these are obviously where the majority (71%) of the costs are. Therefore, this must get the most attention in the manufacturing roadmap.

Source: Greentech Media

Figure 13.1
Cost breakdown for c-Si

The room for maneuvering, however, may be narrow. The materials used in silicon solar cells—silicon, glass, plastic, and aluminum—have all been produced at scale for decades. There is little left to squeeze, but efforts are still ongoing in several areas.

Silicon is still the largest single cost. The main focus for cutting silicon costs has been on the feedstock, the cost of the polysilicon that goes into silicon wafers. There are also efforts to use less silicon by improving (or eliminating) the sawing process, and by making thinner wafers.

Polysilicon costs are expected to stabilize at $20 to $24/kg, up from the unsustainably low price of around $14/kg found in late 2013. Work is being done by SunEdison (formerly MEMC) and others to reduce poly costs further by using a technology called FBR, or Fluidized Bed Reactors. FBRs might help, but they have been around for a long time without yet seeing widespread adoption. Still, there is the chance that FBRs will reduce poly prices another 20% to 30%, and if this technology is very successful poly could get to $15/kg.[37] Still, since the sustainable price for poly is likely higher than current prices, poly is unlikely to be a source of cost reduction for the industry.

An alternative is to just use less silicon. One idea is to use thinner diamond saws to reduce waste (kerf) when slicing wafers from the silicon ingots. Thinner wafers will reduce silicon costs, but it will be difficult to produce them reliably; wafers are already very fragile and subject to breakage. Thinner wafers may also create a loss of efficiency, as less sunlight is absorbed by the thin material.

Another idea is to not saw wafers at all (go "kerf-less") by using cast, deposited, or lift-off wafers that have no kerf loss. But there are two problems with kerf-less techniques: First, they don't save much money, because the processes add complexity that tends to offset any savings. Second, they reduce efficiency and thus could actually increase the final $/watt.

Although silicon is the most expensive single material, it still comprises less than 30% of total materials cost. Major non-silicon material costs include glass, plastic (encapsulant), steel, aluminum, and silver; all are established industries, producing at scale. Most cost-cutting efforts revolve around reducing volumes or eliminating/replacing expensive components. For example, eliminating the frame will save costs as long as an acceptable mounting method is developed. Thin film products have done just that; silicon can copy those techniques.

Silver, which is used to print the grid structure on the solar cell, is expensive; in fact, demand from solar has been driving up the price. Many companies are looking at ways to reduce the quantity of silver required, or to replace it with copper, which is much cheaper. Those efforts should be successful. Even so, these areas are just too small to cut costs by half. Reductions in labor and overhead are required as well.

COGS for Silicon – Labor and Overhead

Labor accounts for 13% of COGS, a relatively high figure compared to mature industries where automation is more common. Overhead comes in at 11% (depreciation of 6% and utilities of 5%), which is relatively low. That makes sense: Lack of automation has yielded low overhead, but high labor cost.

One strategy for silicon is to add automation, exchanging higher depreciation (with more capital investment required) for lower labor costs. Automation also improves process control, so both quality and yield should increase (since human labor is more prone to error). The bottom line: With automation, depreciation might go up 1% to 2%. But the labor costs should go down 5% to 6%, and yield and quality will improve, further reducing costs. The ROI on automation is usually very high.

Putting all of the COGS elements together (materials, labor, and overhead), the picture is not too rosy. Basically, most of the costs already have been wrung out in silicon technology. A likely trajectory for mc-Si is a slow cost reduction of roughly 2-4% year. This is shown in the red line ($/m^2) in Figure 13.2.

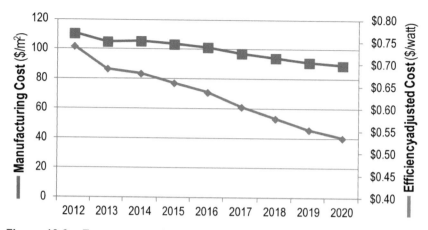

Figure 13.2 – Forecast, mc-Si manufacturing and efficiency-adjusted cost

Costs for Silicon – Efficiency and $/watt

If the potential for reduced manufacturing costs is minimal, can efficiency save the day by improving energy output at the same cost? By combining the forecasted efficiency from Chapter 12 with the manufacturing cost in Figure 13.2, a $/watt forecast is derived (the blue line). Clearly efficiency helps. The $/watt forecast comes down faster than manufacturing cost, because slow but sure efficiency improvements reduce the cost/watt.

This forecast suggests the real cost of mc-Si panels is much higher than many imagine; that mc-Si costs could get below $0.60/watt, but it is not clear the technology will achieve $0.50/watt. This is an important conclusion, because the SunShot grid parity goal is a sales price of $0.50/watt, which would require a manufacturing cost less than $0.50 in order to be sustainable (at least a minimal profit). If this result is surprising, please remember we are being as rigorous as possible, drawing on the work of MIT, Stanford, NREL, Navigant Consulting, and industry sources. We are not using figures from press releases, solar industry boosters, or from China, where much of the reporting is "cash accounting" and not actual cost numbers. For third-party verification of this forecast, please refer to the NREL data provided in Figure 13.3.[38] NREL has performed a rigorous, bottoms-up analysis using raw data to determine the cost of each step in the manufacturing process.

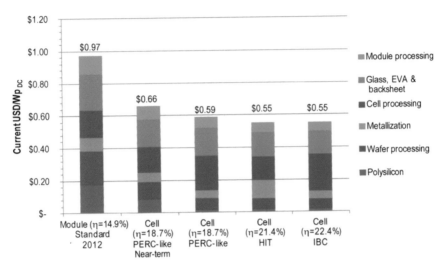

Figure 13.3 – NREL c-Si cost forecast

Note: *NREL assumes profit at each link in the value chain. Actual profit can be zero or negative; therefore, real numbers might be less than NREL's "sustainable" cost estimates. But NREL's assumption is not without merit. For an industry to survive, let alone grow at high rate, a minimum profit is needed in the supply chain.*

In Figure 13.3, in 2012 the standard silicon cells had 14.9% efficiency, and cost of $0.98/watt. NREL envisions a sequence of technologies that reduce costs to $0.56/watt, but there is no time estimate for these improvements. The only "near-term" projection is the second column, which shows 18.7% efficiency and roughly $0.67/watt. But 18.7% in production is years away. Looking further down the curve, there are many technical advances needed to get to $0.56/watt, including achieving 22.4% efficiency. Using 30 years of history as a guide, this could be more than a decade away. Clearly, this third-party work confirms that it may be very difficult to reach the SunShot goal of a sales price of $0.50/watt.

Part 2
COGS for Thin Film – Materials

CdTe and CIGS do not use silicon, so their cost profiles look very different from c-Si. Figure 13.4 shows estimated manufacturing cost percentages for First Solar (representing CdTe) and MiaSolé (representing CIGS), as reported by Greentech Media. First Solar is a good proxy for CdTe, as the company accounts for more than 90% of the world's CdTe capacity. MiaSolé represents less than 5% of the worldwide CIGS capacity...but we have their numbers. The industry leader in CIGS, Solar Frontier, is a division of a large company and its cost data is not available.

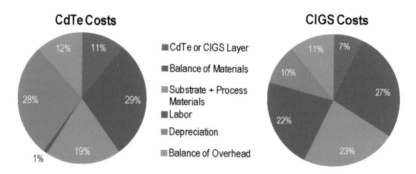

Figure 13.4– Representative CdTe and CIGS costs

Notice that the materials costs (dark blue, red, and green) are still the largest segment of the COGS, at 59% and 57% for CdTe and CIGS, respectively, but lower than for c-Si (71%). Overhead costs are much higher, however – 40% for CdTe and 21% for CIGS, vs. only 11% for c-Si. Labor costs also differ – a very low 1% for CdTe and 22% for CIGS, compared to 13% for c-Si. The CIGS labor costs may be an anomaly based on MiaSolé's singulated cell architecture and small scale (20X smaller than First Solar). At scale with MLI architecture, there is no reason why CIGS labor costs should not be similar to those of CdTe.

At first blush then, it looks like materials have the greatest potential for cost reductions. But the largest component (the red - 29% and 27%) is "balance of materials" – the same glass, plastic, steel, and aluminum that the silicon guys use. These materials are already at scale and cost reduced. The next largest materials cost is the "substrate and process materials" category. For CdTe, this is mostly the glass substrate, and for CIGS, it is stainless steel. In neither case are these materials, also at scale, likely to go much lower in cost. The final category is the one that manufacturers have the most influence over. This is the absorber material, CdTe or CIGS, which absorbs sunlight and converts it to electricity. This cost can be reduced through thinner layers (using less) and more efficient deposition techniques (wasting less). Unfortunately, this is the smallest of the materials cost groups and these processes are also fairly optimized, so there just isn't much room for further reduction.

COGS for Thin Film – Labor and Overhead

Labor and overhead are key to reducing COGS for thin film, accounting for 41% and 43% of total costs, respectively, for CdTe and CIGS (Fig. 13.5). Better equipment design could cut these numbers significantly.

The importance of equipment to the COGS equation is underappreciated (after 25 years in the equipment business, I admit to a bias here). Equipment affects performance on the "watts" side of the $/watt metric, but also effect the "$" side through depreciation, labor, overhead, yield, and even materials cost.

Equipment's most direct impact on costs is through depreciation. Depreciation is a significant part of the costs for CdTe (28%) and CIGS (10%), and the largest component of depreciation by far is the manufacturing equipment. Equipment also has an indirect, but significant, impact on labor and overhead costs.

To illustrate the equipment impact, let's use a next generation thin film factory and compare it to today's standard c-Si factory. A typical Solar 1.0 c-Si production line produces about 25MW of solar panels, and takes up about 1,200 square meters using established silicon process tools. On the other hand, a future thin film production line designed by Siva Power uses about 3,600m^2. This is 3X larger than the silicon line, but…the output is 300MW, or **12 times** as much. The difference in these factories is equipment design. Whereas the c-Si line must process and handle wafers that are only 0.024 m^2 in size, the Siva line can process 2.0m^2 glass plates, or 80 times as large. The high throughput achieved with these large substrates leads to a huge reduction in depreciation, overhead, and labor.

Depreciation: With over 40 years of scaling experience in semiconductors and the FPD industry, the depreciation impact is well-known. Each doubling of factory output only increases factory cost by 20-30%. As a result, the depreciation of a 300MW line is roughly one-third of a 25MW line. That is an incredible 67% reduction in depreciation expense.

Overhead: Many overhead costs are related to the size of the factory. If a high throughput factory with 12X the output only has 3X the facility size, the "floorspace efficiency" is 4X higher; therefore, the overhead should also be much lower. Since some things do not scale linearly with size, the estimated overhead cost of the 300MW line (with very detailed modeling) would be one-third of the c-Si line, a 67% reduction in cost.

Labor: Labor also scales up with the size of the factory and the number of production tools. Because this advanced 300MW factory uses almost 100% automated tools, it would use even less labor, proportionately, than the 25MW silicon factory. The high-throughput line in this model has one-fifth the labor cost of the c-Si factory. That's an 80% reduction in labor cost.

So, for this next generation 300MW line we have a 67% drop in depreciation, another 67% reduction in overhead, and an amazing 80% reduction in labor costs. Sound too good to be true? No…there is precedent for this impact. High throughput equipment is one of the secret ingredients that enabled Moore's Law in the semiconductor industry; a big part of why we are running around with iPhones and IPads. The same impact on the flat-panel display industry is why we have HDTVs, and it is hard to find one of those old "tube" TV's, except maybe in a museum. But equipment has not yet had as big an impact on solar. What's going on?

It's quite simple. Solar is a young industry, and no one has scaled equipment to this level yet. Growth has been accomplished by merely replicating small production lines. One attempt at "high output" scaling was roll-to-roll equipment, but that didn't accomplish the goal either as most R2R lines have had a capacity of less than 25MWs.

R2R – Fact or Fiction?

In the late 2000s, hundreds of millions of dollars flowed into roll-to-roll (R2R) companies like Nanosolar, MiaSolé, Unisolar, Solarion, Global Solar, SoloPower, Ascent, Solexant, Reel Solar, and NuvoSun. Probably because almost every major industry that involves a large, flat area, such as paper, textiles, steel, and plastic, went to R2R equipment to reduce costs. But today most of these solar R2R companies are gone. I didn't understand why R2R didn't work until joining Solexant in 2010.

Solexant had a cool nanoparticle printing technology using R2R equipment. Great idea? Maybe not. When visiting the manufacturing line to watch the equipment work I said, "OK…turn it on." The response came, "It is on." The picture in my mind was a newspaper printing press with the paper flying by so quickly it could slice you in half. So I said, "No, it's not on." The response was, "Yes it is!"

It turns out that Solexant's roll-to-roll speed was not measured in meters per second like other industries, but in meters per minute — 0.2m/min., to be exact. Other industries adopted R2R for speed, not to sound cool. In solar, not a single R2R company ever achieved high speed. For R2R to succeed, it needs to get faster, fast. That has yet to happen.

So equipment scale has a huge impact on depreciation, overhead, and labor, and on capital intensity as well. It also has a more subtle impact on materials costs. Both the amount and quality of feedstock needed to make the CdTe or CIGS absorber are determined by the equipment used. The combined effect of a 2-10X variance in feedstock cost and 30%- 90% variance in feedstock quantity results in a potential 10-30X variance in the materials cost of an absorber.

When adding up the potential savings in materials, labor, overhead, and depreciation, there appear to be more opportunities for cost reduction in thin film than in c-Si. This makes sense because silicon is a much more mature industry; thin film is really just getting started.

Costs for Thin Film – Efficiency and $/watt
As we did for the silicon cost forecast, we will combine the efficiency forecast of the previous chapter with a manufacturing cost forecast to finally arrive at the $/watt cost outlook.

On the manufacturing costs forecast (Fig 13.5) CIGS has trailed CdTe, but the gap is narrowing. Thanks to First Solar CdTe is already at a 2GW scale, and like c-Si, most of the costs have been driven out.

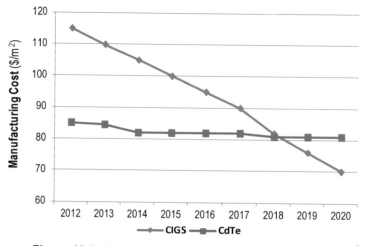

Figure 13.5 – Forecast thin film manufacturing cost - $/m²

The CIGS process is less mature, and no one has achieved economies of scale yet. Because the CdTe and CIGS processes are similar, there is no reason why they should not have similar costs. CIGS should approach CdTe on manufacturing costs within five years. Whereas CdTe seems to have standardized on a small substrate (0.7m²), some manufacturers are looking at depositing CIGS on larger 1.6-2.0m² substrates. In other words, CIGS might be implemented on a bigger, faster production line. If this occurs CIGS will probably surpass CdTe on manufacturing cost.

The efficiency and cost projections can now be combined to derive the $/watt forecast (Fig. 13.6). CdTe maintains its low-cost leadership for several years, but CIGS, with both higher efficiencies and potentially lower manufacturing cost, should catch up and beat CdTe by the end of the decade. These are both very competitive technologies, so it is likely they will share the thin film market.

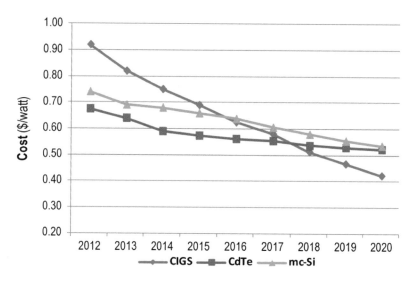

Figure 13.6 – Forecast thin film costs - $/watt

By 2017, both thin film technologies should cost less than mc-Si. They are all so close, though, that other factors may come into play. Reliability, performance in hot environments, or under low or diffuse light conditions, and degradation—all of which impact the final LCOE—will also be factors in which technology prevails. Silicon has a huge lead in market share, so thin film has a challenge to catch up.

Manufacturing Roadmap Challenges

Achieving low-cost manufacturing is not just about reducing COGS. Changes need to be made in manufacturing practices, at least in the United States and especially in solar, to become more competitive. There are three major issues—metrics, best practices, and labor.

The use of **metrics** is one area of management where solar can improve greatly. It is amazing to see how the simple act of posting a chart on a wall seems to make something happen. There have been two fundamental flaws in the use of metrics in solar: 1) few or weak metrics, and 2) using the wrong ones. Specifically, the industry has been too single-minded about efficiency, and needs to balance that metric with a stronger cost metric, COO.

COO

Re-visiting the cost equation, as we have seen, the $/watt breaks down into just two components:

$$\$/watt = \$/m^2 \div watts/m^2$$

Watts/m^2 (the efficiency) and $/$m^2$ (the manufacturing cost) are both important, but the solar industry must focus more on the cost side. Other industries, like semiconductors, went through this transition from technology to cost focus long ago. Solar can and should benefit from that experience, adopting an analytical tool used for decades to measure, manage, and predict the manufacturing cost; that $/$m^2$ tool is the Cost of Ownership (COO).

COO includes everything. Such boring factory cost issues as equipment consumables, spare parts, the cost of electricity, water and gas usage, and waste disposal all need to be factored into the cost of manufacturing solar panels. If companies do not measure these in order to learn where the high and low cost points are, cost-reduction efforts will be misdirected.

> ### COO Emerges in Semiconductors
>
> *When I worked as a product manager at Applied Materials in the 1980s, we were losing sales because our competitor's product was cheaper. But if all costs (not just initial sales price) were factored in, Applied should have won. Looking for a way to represent that, I developed an equation that factored in all of the costs to own and operate a system, and compared that to the system output. This spreadsheet represented the true cost per unit ($/wafer) to own and operate the system. When the customer saw this cost of ownership (COO) data, Applied started winning more sales.*
>
> *When I founded Novellus, we calculated the COO before we built the prototype, in order to predict cost competitiveness. It worked. Our first product, based on the COO model, was a smashing success. Soon, other companies and our customers figured out how powerful the model was, and COO became a standard tool in the highly cost-sensitive semiconductor industry. Today, the DOE has included the COO metric in their Manufacturing Roadmap for Solid State Lighting. Hopefully, COO will eventually find widespread adoption in solar as well.*

R&D can also be misdirected if the costs are not better understood. The highest efficiency devices, like those made by Sanyo, SunPower, or Alta Devices, rate up to 30% higher on performance, but cost 100% to 800% more.

Figure 13.7 illustrates how things can go wrong. It shows projected efficiency improvements, along with projected cost increases of some of the more popular technologies. The "Net Benefit" column calculates the outcomes, in $/watt. The "Efficiency Adjusted" column factors in the BOS value of higher efficiency, valuing efficiency improvements 3X more than costs. These are numbers for illustration purposes, but if they are even close to reality, the large areas of red tell a lesson: "Not all exciting new technologies are cost-effective." Most of these technologies have yet to find their way into high-volume production, suggesting that the Net Benefit column might be close to the truth.

Silicon Technology	Efficiency (watt/m^2)	Cost ($/m^2)	Net Benefit ($/watt)	Eff. Adj. ($/watt)
Monocast	5%	15%	-10%	-0-
HIT Technology	15%	60%	-39%	-10%
Selective Emitter	6%	8%	-2%	8%
Double Print	4%	8%	-4%	4%
Kerf-less Thin Si	-15%	-10%	-6%	-64%
Hi-eff Lift-off Silicon	10%	40%	-27%	-8%
Thin Film Technology				
Metal Wrap Through	8%	30%	-20%	-5%
Dual-Junction	15%	50%	-30%	-3%
High-Efficiency III-V	70%	600%	-253%	-186%
NanoParticle Printing	-10%	-5%	-6%	-36%
Nano-structures	-20%	30%	-63%	-225%
R2R processing (vs. MLI)	-4%	10%	-15%	-25%

Figure 13.7 – Illustrative example - efficiency vs. cost increases

The point is that COO can help to determine the net benefit without investing $100 million to figure it out the hard way. Siva Power used COO to discard many of the technologies listed above, and to determine which options are viable for Solar 2.0.

Another best practice needed in solar is the smooth and fast transfer of technology from R&D into manufacturing—in other words, from lab to fab. Part of the solution is to scale up at the right time. As much as 80% to 90% of the technical and market risk should be gone before scaling. Companies that do not do that—think Solyndra or Nanosolar—just burn through money. Those two companies went through $1 billion because they scaled before basic cost and technical milestones were met. Even after initial fab ramp up, an effective process to continually bring innovations from R&D into high-volume manufacturing is essential.

Labor, defined as the quantity, quality, and cost of the workforce, is another challenge. In terms of quantity, it can be difficult in some countries or regions to find enough people with the skills and willingness to work in a solar factory. Therefore, it's important to locate in an area where the availability and aptitude of the workforce matches the needs of the company. Workforce cost is a major issue, of course. This does not mean, for example, that you cannot manufacture in the United States, but it might mean that you can't manufacture in the high cost environment of Silicon Valley. There are many countries around the world, and states in the US, with strong manufacturing backgrounds and apprentice systems. Naturally, if those regions also feature lower labor costs, that can be helpful, too.

Think of the adage "work smarter, not harder." In this case, smarter might mean fully automated equipment. In the case of the 300MW production line for thin film mentioned earlier, high-speed automation cut labor costs by 80%. The kind of labor required also changed—fewer operators and more skilled technicians and engineers. With this kind of advanced manufacturing technology, the need for a better educated, highly skilled workforce (not necessarily low-cost labor), suggests a competitive advantage for countries like Germany and the US.

~ ~ ~ ~ ~ ~ ~ ~ ~ ~ ~

The drive for cost reduction is a defining feature of Solar 2.0. Every cost element must be understood, analyzed, and reduced. Many of the costs of silicon panel manufacturing have already been rung out. But since thin film is much less mature, there is much more room for cost reduction. Thin film technologies could be the first to reach $0.40/watt, meeting the SunShot goal of a sustainable selling price of $0.50/watt.

Chapter 14

Product Roadmap

Technically, solar power plants are fairly simple. Panels are roughly the same, whether they are for a small 2kW home system or a 200MW utility system; they are strung together and mounted on a roof or on the ground. Wiring, cables and connectors knit the panels together to collect the electricity. Then an inverter converts the DC power from the panels into AC electricity so that it can be connected to the grid or used directly in homes, offices, or factories. In the near future, two trends are likely. First, these panels, components, and systems will become optimized for each market segment: residential, commercial, and utility-scale. Second, as systems thinking begins to shape design, these parts will all evolve. Innovation will play an important role in the optimization and evolution of solar products and systems.

The solar industry has long tended to focus on technology, and more recently on cost-cutting. In both cases, the emphasis has been on individual components as opposed to the system. With the commoditization of the industry, differentiation is becoming the key to profitability. There is an opportunity for innovation on the system side that will provide this differentiation.

Figure 14.1 shows a typical solar power system and all of its components. The concept is very simple: Solar panels are mounted on a structure (such as a roof), interconnected electrically, and then connected to the grid (utility) through an inverter.

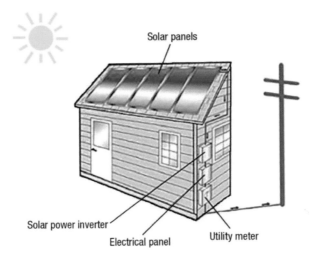

Figure 14.1 – Typical grid-tied solar power system

The three primary markets for electricity (utility scale, commercial, and residential) each have different needs, and systems will likely evolve to meet them. In this chapter, we look at each market segment, and break down the system cost to see where the best opportunities for cost reduction exist. Since panel costs were addressed in Chapter 13, here the focus will be on BOS costs.

Utility scale: Figure 14.2 compares the 2011 average cost breakdown for two utility-scale (10MW) projects—one a multi c-Si system at 14.4% efficiency, and the other a thin-film CdTe system at 11.5% efficiency.

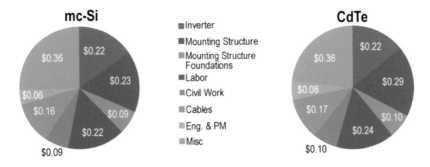

Figure 14.2 – BOS cost breakdown for US mc-Si and CdTe projects

The two are pretty similar, with the largest cost component (after the panels) being the mounting structure, followed by labor and the inverter.[39] Hardware and non-hardware costs are almost evenly split. To reduce the cost of mounting, an integrated system could be designed that combines the physical mounting and electrical interconnect. This kind of innovation would address not just hardware costs, but could simplify installation therefore reducing labor costs as well.

One innovation for these ground-mounted systems is in the footing, which often involves putting a concrete foundation in the ground to hold the racking (mounting system). A simplified footer eliminates concrete, reducing both cost and installation time. Without waiting for a concrete footing to cure, a crew can install footers and the rack in one operation, saving time and labor.

Another area of interesting systems innovation has to do with scale. On a 100MW utility-scale project, the current practice is to transport and then individually bolt, hand-wire, and install each individual panel; this could be 500,000 panels at 200W each. The number of individual operations is immense, as is the margin for error.

There are a couple of ways to simplify this process and reduce labor costs, both involving larger-scale components. The first option is to pre-assemble the strings in the factory. The pre-assembled strings would contain integrated wiring, and would mount in the field to the racking system using snap-fit or friction couplings, enabling automation of the entire process. All of this could be done using robotic systems under GPS control. Although the capital cost would be higher, the savings in labor and overhead would more than compensate.

The second idea is to incorporate both racking and wiring as integral parts of what might be called a "megapanel." The goal is in-field deployment of integrated, large-area modules, say 5 to 10 kW.

The factory-built megapanels would be loaded on flatbeds and placed right on the just-installed footings.

Both of these approaches would cut labor costs by reducing the extent of field work, which is much more expensive than in-factory labor. The cost of mounting and electrical connections would also be reduced because there would be fewer individual pieces.

Commercial: Figure 14.3 provides a breakdown of the BOS costs for a 2011 commercial-scale (500kW) installation in the American Midwest.[40]

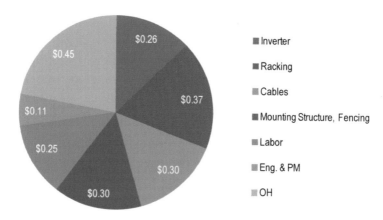

Figure 14.3 – BOS costs for a commercial installation

Again, the racking is the most expensive component, followed by miscellaneous electrical (wiring, combiner, system monitor, conduit, fuses, and breakers) and the inverter. Hardware is about 36% of the total, but 64% is non-hardware cost, with profit and labor being the largest components. Soft costs are now more important than hardware.

Commercial rooftops have the potential to be a big part of the distributed energy solution in Solar 2.0, so addressing these challenges should be a high priority. A good place to start is panel architecture. Figure 14.4 shows a unique mounting system that not only provides a tilt angle to maximize harvest of direct sunlight, but also includes a low-cost reflector in the space between panels.

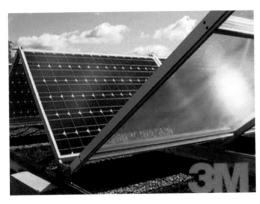

Figure 14.4 – Novel panel architecture

The reflector directs light back to the panel and improves efficiency. This is a clever product design that could help the commercial rooftop market. It does not, of course, address soft costs.

Solyndra had a great product to reduce soft costs for commercial installations. They went bust, as we know, but in that process they provided some good examples of what to do—and what not to do.

What did Solyndra do right? Their system-design mentality for a commercial application was dead-on. The system was easy to install, without ballasts or troublesome roof penetrations (the anchoring devices required to hold the solar panels down in high-wind conditions). The company also took advantage of the fact that some roofs have a white coating that would reflect sunlight, and devised a way to harvest this indirect light. Finally, it made the modules easy to assemble, with the electrical connections being an automated part of the assembly process. These are all promising solutions.

Where did Solyndra go wrong? Remember our motto: Cost is everything. The solutions were simply too expensive, and management failed to face that truth. Every innovation that Solyndra developed to reduce BOS costs worked, but at too high a price. That doesn't mean others can't figure out a way to do it cheaply. The ideas Solyndra pioneered are a good template for a commercial roadmap for Solar 2.0:

1. Mounting that does not require roof penetrations or bal- lasts
2. Easy-to-install systems that reduce labor costs
3. Automatic wiring assembly to reduce wiring hardware and labor costs
4. Methods to harvest reflected or diffuse light, and thus improve efficiency

Residential: The cost structure in the residential market is even more dominated by soft costs. In the US especially, costs such as customer acquisition, installation, labor, permitting, inspection, in- terconnection, and financing comprise more than two-thirds (69%) of the expense of residential installations—more than twice as high as in Germany (Fig. 14.5). [41]

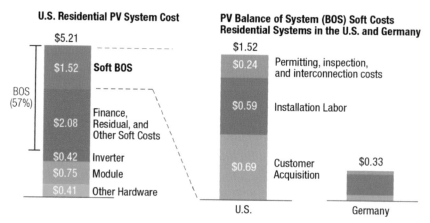

Figure 14.5 – BOS cost for a residential system, US vs. Germany

Clearly, there is enormous opportunity to improve the economics of US residential solar by reducing soft costs. Chapters 15 and 17 go into detail on business model and policy recommendations, but here are a few quick ideas:

• Establish standard system designs

- Allow a nation-wide, one-day permit process for such standard systems
- Establish feed-in-tariffs, eliminating the need for complex financing schemes
- Change from the US 600 Vdc to the European 1000-1500 Vdc limit; this would allow longer system strings, significantly reducing BOS costs

These ideas are great, but there are system and product designs that could reduce or eliminate soft costs as well. Product innovation should play a bigger role in Solar 2.0—and it is long overdue. Over the past 30 years many things have changed about the industry, but solar panels have remained basically the same (other than becoming cheaper). For the residential market, new products could include:

- Integrated, self-installing "sloped roof" mounting systems
- Economical solar shingles that integrate solar panels with roofing tiles
- Modular AC panels that eliminate inverters, and enable simpler, scalable systems

Then there is the product opportunity that could be a game-changer—the "total solution," or the "solar appliance." This doesn't exist yet, but some pieces do. Imagine if you could just go down to Home Depot or Best Buy and buy a solar power system, just like you'd buy a refrigerator, washing machine, dryer, or HDTV. You wouldn't need a permit or a home inspection for any of those "appliances." There would be almost no soft costs. You wouldn't need permission to get an interconnection, and you wouldn't be paying for an expensive sales team.

This would be a dream come true. Such a solar appliance would significantly reduce the barriers to acquiring solar power, and empower the American homeowner to be more energy-independent at a personal level. That's the dream. The reality is that there are a number of hairy challenges to make the solar appliance happen:

1. It requires new hardware. This panel will be different from current designs. It will need micro-inverters, so that each panel becomes an AC module. This makes system expansion easier. Micro-inverters exist, but current offerings are still too expensive and not reliable enough. Also, the hardware needs to have a simple modular mounting system for easy installation and expansion.

2. Homes need to be pre-wired. Just as with your washing machine or dryer, the home needs to be ready to accept a solar appliance. Until building codes for new construction incorporate such features, perhaps a retrofit kit could be developed to make it easy to wire up. This is where the standards mentioned above would help. A standard system could be pre-approved, UL-certified and ready to go without permitting.

3. Utility companies need to get on board. For obvious reasons, utilities don't necessarily want their customers to have independent energy sources. Regulators, too, can be skeptical, seeing this idea as unsafe, unreliable, and hard to manage. To overcome these concerns, it is crucial that the solar product is designed, certified safe, and installed properly.

4. The DIY (do-it-yourself) community would need to adopt the process. DIY-ers need a simple installation process. For those unwilling or unable to get up on a roof, a lot of installers will be needed. If Home Depot, Best Buy, and others got into the act with their extensive network of stores and installers, it would create a great channel for the solar appliance.

There has been some progress on item one, the product itself. Startup companies like Armageddon have developed a solar appliance product; it's still too expensive, but it's an intriguing start. The real limitations are items two and three which create obstacles to a simple plug-and-play system. It will probably require us, the consumers, to demand the needed changes in permitting and building codes.

~ ~ ~ ~ ~ ~ ~ ~ ~ ~ ~ ~ ~

Product innovation can have a big impact in every sector of the industry, whether be it utility-scale, commercial or residential. The solar product evolution will be fun to watch. Personally, I can't wait to see the solar appliance come to a store near me.

Chapter 15

Business Model Roadmap

> *Silicon Valley is known for technical innovation, and Solar 2.0 certainly needs this kind of innovation. But as soft costs become more important, solar also needs market innovation. There is a huge opportunity to accelerate solar deployment through novel financing arrangements, more efficient sales processes, and improved distribution and installation methods. Can this work? SolarCity proves it can: On the basis of a new business model, it has become one of the most valuable solar companies in the world. New business models will rival technology innovation for impact on the solar industry.*

In 2012, Rob Day of Black Coral Capital said something important about cleantech that was also pertinent to solar:[42] "The next wave won't be tech-oriented. It will be commerce-oriented and application-driven." Rob started the discussion on "market innovation" — a term that includes new business models, more efficient ways to acquire customers, novel financing methods, and ways to reduce soft costs that do not involve technology. For solar to reach its full potential, the focus needs to shift away from technology and toward deployment. This chapter will explore how that can happen, first for the upstream sector, then for downstream and finally for financing.

Upstream Business Models

Silicon: For solar to progress from the megawatt to the gigawatt era, a great deal more global capacity will be needed. For silicon solar companies, this transition will be characterized by consolidation and vertical integration (see Chapter 9). With their large silicon manufacturing base Chinese companies are likely to continue to dominate the field, but manufacturing is likely to globalize, as has occurred in other industries. Solar will evolve to a "glocal" business model—globally managed, but with local manufacturing in each major market. Why?

As solar becomes the renewable energy of the future, its strategic value will create a case for policies such as incentives, trade protection, and regulations to require local manufacturing. There is precedent for this in several industries, such as auto and semiconductors. No major economy wants to be entirely dependent on other nations for a resource as strategic as energy. So it is likely that there will be political action in countries around the world to put local content rules in place, providing a strong incentive for local manufacturing. This has already happened in India. Most of the top 10 solar companies will probably have factories in all three major markets: Asia, Europe, and the Americas.

Another reason to move closer to the end markets is cost. Companies will scrutinize every penny spent. Solar panels are relatively heavy and bulky, which means high shipping costs. That is an incentive to manufacture closer to the consumer and the supply chain. One approach to consider is to take a page out of the flat panel display (FPD) industry playbook and "co-locate," eliminating incoming shipping costs altogether. Glass is the heaviest and bulkiest solar component. FPD companies have been huge consumers of glass for years, and as a result, modern FPD factories usually co-locate with an existing glass manufacturing plant. All they need is a forklift to move the glass from one building to the next. Solar is likely to do the same.

Thin film: Compared to silicon, thin film is an immature industry. Unfortunately, many thin film companies tried to follow the silicon business model, buying equipment and materials from an established supply chain that piggybacked on the silicon semiconductor industry. But thin film is different, and those suppliers did not have the expertise to adapt their equipment to the unique needs of the thin film industry. That played a role in the thin film bloodbath, which saw many companies go under or bankrupt.

The correct business model for thin film is to develop core competencies (Fig. 15.1.) in materials technologies, equipment, and solar devices (cells and panels). This is a tough model to follow, as entire industries have grown up around these areas, and it's hard enough developing one technology, let alone three. Few will be able to bring together this three-fold expertise, and the barriers to entry will be high.

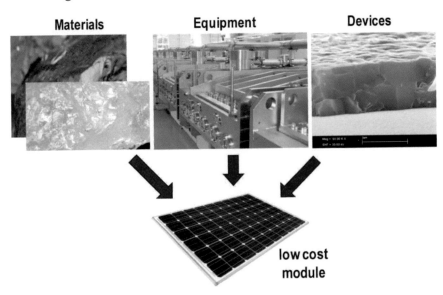

Figure 15.1 – Proposed thin film business model

But because of these barriers to entry, the number of competitors will be lower and profitability will be higher. The recent experience of First Solar makes the case: They pioneered and proved this integrated business model (of materials, equipment, and devices) and still benefits from it. It is an awesome competitive advantage.

The problem is that the materials, equipment, and device business-es attract different kinds of people, and require different skills. To execute this more complex business model, a successful thin film company needs three leaders. Three CEOs might be difficult (to say the least). One way to accomplish the same thing would be to hire a "dream team" of strong C-level (CEO, COO, or CTO) executives, with skills in each of the required disciplines.

Materials and Equipment: The materials segment is already mature and competitive, so the business model is set. The equipment sector is a different story. There are two competing business models for equipment. In one case, turnkey factories are designed by a single equipment supplier with a limited tool set. Alternatively, "best-of-breed," operations are set up, in which the factory designer identi-fies the best tool for each process and integrates them into a cus-tom-manufacturing operation. Then management can optimize each step while minimally affecting the others. This enabling of continuous improvement of each manufacturing step is what led to the incredible gains in the IC, LED, and FPD industries. It is part of what made Moore's Law work. In an industry where technological change is constant, the best-of-breed strategy allows a manufacturer to achieve the highest performance at the lowest possible cost. This gives it a big advantage over turnkey factories, which will likely die, as Applied Materials' SunFab did in 2010, and Oerlikon is do-ing in 2014.

The silicon equipment supply chain is mature and firmly in place. The best-of-breed model has won, and c-Si manufacturers have a menu of quality suppliers to choose from. Also, there are many sili-con solar companies, so equipment suppliers have many customers to serve, which diversifies their sales risk. It is sometimes tumultu-ous (due to cyclicality), but an otherwise stable ecosystem.

Thin film, on the other hand, is in flux. With many thin film manu-facturers exiting or consolidating, the industry could find itself with only five or so left; this will not support an independent thin film equipment industry. There is another problem: Every thin film company has a unique technology that requires a different equipment set. There are at least five ways to deposit the absorber layer, as well as three or four substrate materials. There are also completely different tool designs such as batch and continuous roll-to-roll. With this diversity and so few customers, no equipment company will achieve enough volume to make product lines viable.

Instead, successful thin film companies will emulate First Solar and buy or make highly specialized, custom equipment, keeping their technology close to their vest. The equipment business model in thin film will either be vertical integration or very close, exclusive relationships between the equipment supplier and the thin film company.

Downstream Business Models

The downstream market is the most exciting, dynamic, and fastest-growing segment of the solar industry. It is also very complicated, including components like:

- System builders - EPC (Engineer, Procure, and Construct)
- Project developers - manage everything including EPC
- Financiers – provide the money
- Operators – operate the solar power system once it's built
- Customers (or off-takers) – use or buy the electricity

Another downstream complication is that the optimal business models are different for the three end users: utility, commercial and residential. And, of course, the scale varies wildly.

End users	Project Size (kW)
Residential	2 - 20
Commercial-Industrial	10 - 1,000
Utility-scale	10,000 - 1,000,000

Residential systems can be so small (often just 8 to 10 solar panels), that there may be no value for an EPC company or financial institution to get involved. The complexity here is in acquiring and managing a large number of customers while keeping costs down. So-larCity, SunEdison, Sungevity, and others are seeking to do just this, through a TPO (third-party owned) business model in which they take care of everything from design and installation to operations and maintenance.

In fact, TPO is becoming the dominant downstream business model for the residential market. A key component is the financing strategy. The solar power system is owned by a third party and leased to the residential consumer. The homeowner, therefore, does not have to pay the high upfront cost, which can make going solar prohibitively expensive. With this financing mechanism, more and more people can afford solar. Figure 15.2 shows that since 2009, these third-party systems have gone from less than 10% of solar installations to more than 80% in some states. So the leasing financial strategy is working. The sales strategy? Not so much.

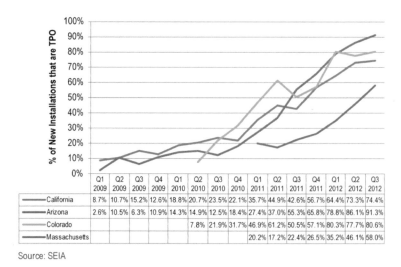

	Q1 2009	Q2 2009	Q3 2009	Q4 2009	Q1 2010	Q2 2010	Q3 2010	Q4 2010	Q1 2011	Q2 2011	Q3 2011	Q4 2011	Q1 2012	Q2 2012	Q3 2012
California	8.7%	10.7%	15.2%	12.6%	18.8%	20.7%	23.5%	22.1%	35.7%	44.9%	42.6%	56.7%	64.4%	73.3%	74.4%
Arizona	2.6%	10.5%	6.3%	10.9%	14.3%	14.9%	12.5%	18.4%	27.4%	37.0%	55.3%	65.8%	78.8%	86.1%	91.3%
Colorado					7.8%	21.9%	31.7%	46.9%	61.2%	50.5%	57.1%	80.3%	77.7%	80.6%	
Massachusetts									20.2%	17.2%	22.4%	26.5%	35.2%	46.1%	58.0%

Source: SEIA

Figure 15.2 – TPO systems as percent of new solar installations

Customer acquisition costs are still too high. Another innovative business model may be needed to solve this problem. One could simply develop more efficient sales methods internally, but perhaps a strategic alliance with an already existing sales channel, like Best Buy or Home Depot, will be the better strategy. These channels may already be selling panels to do-it-yourselfers, but they rarely offer the complete package, including the financing that the TPOs offer. It might be a good combination.

The full DIY model, introduced in Chapter 14, could turn things upside down, obsoleting the TPO model. If the solar appliance described in that chapter were available, consumers could buy a solar power system without a TPO, much as they buy a refrigerator or television. Of course there is still the need for financing, so a DIY model would only pose a threat to TPO if the price was low enough that people could buy it outright, or if financing terms were made available by retailers, much like when buying a car.

Another exciting market innovation to watch is "community solar." According to an NREL study, *A Guide to Community Shared Solar: Utility, Private, and Nonprofit Project Development,* as many as 75% of residential customers may not be good candidates for solar installation.[43] Maybe their roofs face the wrong direction, or there is too much shade, or they are renters. With the shared or community solar business model, the physical installation can be anywhere in the neighborhood, optimally located for good sun and mounting conditions; for example, at a local school, grocery store, library, or any building with a large, flat, easy-to-work-with roof. The electricity is fed into the grid and the shared "owners" get compensated with a reduction of their electricity bill at home, based on their ownership percentage.

This idea has great potential to make the acquisition of solar power easier and cheaper. Aggregating demand creates economies of scale, for both hard and soft costs. Instead of 20 custom designs, 20 permits, and 20 contracts, all of these are done once for 20 households. Community solar could become the business model that makes low-cost solar widely available.

The commercial market is more complicated. The end user of electricity is usually a business that might be leasing the building. The business pays the electricity bill, but is not motivated to invest in a solar power system because it could lose its investment at the end of the lease. Owners don't have much of an incentive either because they don't pay the electricity bill.

There is a huge opportunity for new business models to open up the commercial market. The TPO model can work where the owner is the occupier, and this is being addressed with major retailers like Walmart, Costco, and IKEA already on the solar bandwagon (see Fig. 15.3). In fact, the TPO model might be great here because the one weakness, the customer acquisition costs, should go way down. Not only is the average installation much larger — distributing the cost over more "watts" — but a contract covering hundreds or even thousands of stores could be negotiated at one time with corporate headquarters.

The tougher segment of the owner-occupier piece will be the smaller business, or big chains with smaller roofs, like McDonald's or 7-11 stores. Smaller businesses would see the same high sales costs as homeowners. Also, large chains with small roofs may exacerbate the energy density problem seen in residential — that is, the area of the roof may not allow enough solar panels to meet the energy requirements.

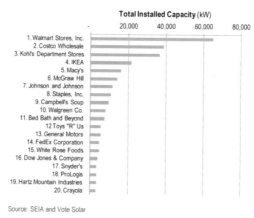

Figure 15.3 – Solar installs Top 20

Still, SolarCity, SunEdison and others are expected to attack this market, as the revenue per sale should be higher than the residential segment because one sales campaign could convert an entire chain.

In the owner-different-from-occupier scenario, the community solar business model might be the solution. Essentially, you leave the occupier out of the equation. Instead, find a long-term customer who wants to buy electricity and then get a nearby building owner to lease its otherwise worthless rooftop separately. By installing a community solar system on this rooftop, all the neighbors become potential customers for the electricity.

As with residential, this would require a TPO. The developer holds the 25-year rooftop lease, and on that basis signs up consumers, thus getting a 25-year income stream from the electricity generated. The economics can work if the utility companies agree to the arrangements, so the customers can buy electricity from the community solar developer and get that amount deducted from their utility bill. Unfortunately, as with all community solar ideas, the utilities are in the middle of the equation. They own the wires that connect all the customers to all the sources of electricity, even community sources. But this model can work: There are at least 31 shared community renewables projects in 12 US states.[44]

Personally, I think community solar is the best solution for both residential and commercial markets. We just need utility companies to cooperate to make it more widely available.

The utility-scale sector is the most complicated market—and it is often controversial. Few would complain when a homeowner puts solar panels on their roof, but some people object to utility-scale projects which often take up wide swaths of public land and may require unsightly transmission lines. Distributed energy is the most promising path to energy security, which means residential and commercial solar. Still, given the need for clean energy, we need every possible source of solar energy. Utility-scale solar should be part of the solution.

A significant utility-scale solar project will often work with separate entities for each downstream function: design, installation, finance, operations, and project management. And there are many different companies involved—typically a developer, an EPC company, a bank (or other financing entity), solar component manufacturers, an operating company, and the end customer, usually a utility.

One positive factor is that the business processes for these huge projects are well established—it's all been done before, as far back as the pyramids (or at least the Hoover Dam and Alaska pipeline). The only area that will need significant change is financing. With billions involved at the utility-scale level, capital assumes a larger role than in smaller systems, and new financing tools are being explored.

In the US, solar is subsidized with a federal tax benefit, and utility-scale projects are typically financed through tax equity. A tax-equity investor helps a project developer monetize the tax benefits, including the Investment Tax Credit (ITC). Other countries have different subsidies, but often the government is involved because utility-scale projects are so big.

New financial tools to utilize tax credits (such as Solar Renewable Energy Credits, or SRECs) are evolving, and have worked to varying degrees. The use of real estate investment trusts (REITs) and master limited partnerships (MLPs) could also help to reduce the cost of capital. These are US-based tools, so I won't go into too much detail. The bottom line is that innovative financing schemes are important to utility-scale solar installations everywhere.

Financial Business Models

There is a significant financial component to solar installations, regardless of the market segment. In fact, as solar reaches grid parity, the main thing stopping rapid deployment will not be costs (as measured over the lifetime of the project), but financing those costs upfront. Clever, new financial models are being tested all the time.

One idea that has many people excited is securitization. Because a solar lease is backed by a fixed, cash-generating asset (i.e., panels generating electricity), it can also lend itself to securitization. "Securitization," explains GTM Research VP Shayle Kann, "is taking a portfolio of contracted revenue from solar projects, bundling it, and selling it as an individual security."[45] Sungevity, among other TPO companies, is now bundling its leases and selling "solar bonds." In 2013, SolarCity announced its first "securitization" product, and in 2014 followed up with a second.

Securitization opens up more financial markets to solar, thereby reducing the cost of capital. Once the leases are bundled, the TPOs can sell them to financial institutions that specialize in buying low-risk, long-term assets. These institutions have a lot of capital, which solar developers would love to access. A ton of data is required to quantify the low-risk part, but once this is established, more capital opportunities will open up, and interest rates will fall. A report from T. Alafita and J. M. Pearce in early 2014[46] stated that securitization could reduce the cost of capital from 5% to 13%.

~ ~ ~ ~ ~ ~ ~ ~ ~ ~ ~ ~

New business models are important for the viability of both silicon and thin film upstream companies. But they are also crucial to the expanding downstream business where solar deployment is accelerating. These innovations will be at the core of reducing those stubborn soft costs across all the sectors: residential, commercial, and utility-scale. Technology and policy have dominated the history of solar, but cost and business models will dominate its future.

Chapter 16

Investment Roadmap

Solar needs capital for both innovation and growth. Un-fortunately, in the difficult market conditions that pre-vailed from 2006-10, many investors left the sector and are staying out. Without capital, the future of the indus-try is at risk. In particular, finding investors who will put money in late-stage development is a major challenge. This chapter argues that there are many solar opportuni-ties where investors can earn healthy profits; it's a matter of knowing the market, evaluating the risks, and learning from recent history. In early 2014, for example, SolarCi-ty's stock price was up more than 600% from its IPO. Ten specific profiles are offered as good investment opportuni-ties.

During my stint in venture capital, I spent nearly two years analyz-ing the solar market and continuously refining an investment thesis for the sector. Later, when I went into the trenches as a solar com-pany CEO, I gained new perspectives on what it really takes to suc-ceed in this very competitive market. In this chapter, I share some of what I've learned — sometimes the hard way — about the oppor-tunities, risks and potential rewards of investing in solar.

Solar investment opportunities fall into several major categories and more than 50 individual segments (Fig. 16.1), each with its own unique characteristics: different market sizes, growth rates, compet-itive forces, headwinds or tailwinds, policy dependence, financial models, margin expectations, and strategic importance. Each of these issues affects the investment-worthiness of the segment. There is also the transformative nature of the investments to consider. Some technologies have a chance to change the industry, and these are exciting for investors looking for a grand slam.

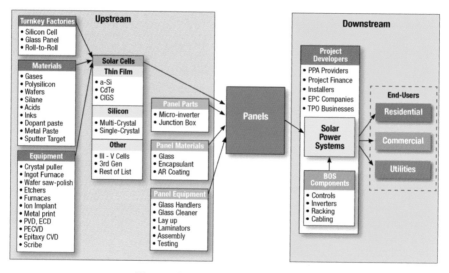

Figure 16.1 – Solar ecosystem

Figure 16.2 looks at eight of the 50 sectors in solar, and evaluates the relevant factors based on impact on return. Each sector is graded on a scale; "three" is most investable, and "one" …not so much.

Sectors - Subsectors	Size	Growth Rate	Entrench Competitor	Margin	Tailwind	Strategic Value	Chance to Transform	Sector Summary
weighting factor	2	1	1	2	1	1	3	
Energy Providers								
PPA suppliers	2	3	3	3	1	3	1	23
Independents - ISOs	2	1	1	2	1	3	1	17
Utilities	3	1	1	1	1	3	1	17
Installers								
3rd Party Installers	3	3	2	1	2	2	3	26
Community Systems	3	1	3	2	3	2	3	28
Residential	3	3	3	1	2	2	1	21
Commercial-Industrial	3	3	2	2	1	2	2	24
Utility scale	3	2	3	1	2	2	1	20

Figure 16.2 – Sample of solar sector evaluation

Note: The spreadsheet shown is not intended to provide specific guidance; it is for illustration purposes only.

The numbers are added up and then coded red, yellow, or green. Green means "most investable," red is "don't touch it with a ten-foot pole," and yellow is "watch and wait."

On the basis of the analysis done, the following are ten promising investment opportunities:

1. Make silicon more profitable: China has cornered the market in c-Si solar manufacturing. One strategy is to not compete with Chinese manufacturers, but to sell to them by addressing their most pressing need: profitability.

One way to enhance profitability is to reduce silicon use. Silicon is the single largest cost with c-Si solar panels. When in the VC world, I looked at about a dozen companies that were trying to cut kerf losses (in fact, for a short time I was CEO of a kerf-less solar company). Some technologies try to cast wafers; others peel off thin layers of silicon from a thicker wafer. Most, however, try putting a thin layer of high-quality silicon on top of a cheaper substrate. This profile became known as SOC, or "silicon on (top of) cheap." But the cheap substrates are hard to work with, and it was expensive to put good quality silicon on top, so most companies failed or abandoned the idea. Cutting silicon with a thinner saw (diamond wire), thereby wasting less silicon, seems to be a viable technology. But most other methods add so many additional costs that the status quo is better. Also, some kerfless designs reduce efficiency. Still, the concept definitely makes the investment watch list, because it addresses the single largest cost issue with c-Si solar panels.

Another possibility is to replace silver, the second-most expensive material in silicon solar panels. There are proposed replacements, like copper, but such changes can affect both performance and cost. One exciting idea here is to replace silver printing with wires that are laminated onto the wafer surface. MiaSolé has successfully implemented this idea on thin film, and the process may find its way to the silicon side. Any workable, low-cost replacement for silver would see rapid adoption.

A third option is to enhance efficiency. To date, most such efforts have not been cost-effective, meaning that the cost increases outweighed the efficiency gains. One of the most popular approaches, "selective emitter," is already in limited production. This process reduces resistive losses where the silver contacts the silicon wafer.

> Just a comment: when looking at any of these new technologies to improve efficiency, one must always assess the cost impact in relationship to the efficiency improvement. The industry does not have a standard method for making cost assessments, so wild cost claims are all too common. As a result, the needed cost vs. performance trade-off is not done or is done with bad data. Using the Cost of Ownership (COO) metric outlined in Chapter 13 would save both time and money. When reviewing business plans as a VC, COO screening eliminated 90% of prospects; the numbers just did not pencil out. One company proposed a plan to achieve $0.50/watt; the actual result was closer to $8.00, more than an order of magnitude higher.

2. Place a bet on thin film: At its peak in 2008, when the solar market was only 6GW, First Solar had a valuation of more than $20 billion. When the Solar 2.0 thin film winner emerges the market will be closer to 60GW, so the valuation of the market leader will probably be much greater, too. Because thin film is an inherently lower-cost technology than silicon, the thin film leader could well become the solar leader. Yes, there is risk in thin film, but the rewards could be huge. If one of the thin film companies listed below achieves just 3% market share, the value would be in the billions. If it would achieve the market share First Solar had in the past, the value would be in the *tens* of billions. This is by far the highest leverage investment on this list.

The market will likely support five thin film winners (see Chapter 11). Two of these, First Solar and Hanergy, already exist. That leaves three spots open to new players. The following are the likely winners:

A second CdTe company: This theoretical challenger to First Solar triples the substrate size, with a process that matches First Solar's speed and simplicity.

CIGS MLI: This combines the best thin film materials, CIGS, with First Solar's proven, lowest-cost manufacturing process, monolithic integration. The possibility of reaching high efficiency at such a low cost is exciting.

R2R CIGS: This is a 2.0 version of MiaSolé. In order for roll-to-roll technology to succeed, the speeds need to increase about 5X, achieving five square meters a minute. This technology could not only compete with the other options, but could open up the flexible, lightweight market.

3. Narrow the efficiency gap and improve yields: There is a large gap between efficiency in the laboratory and efficiency on the factory floor. The solution is better process control, and technologies that enable this could earn excellent returns.

Metrology is the science of measurement; it is crucial to process control and therefore to closing the efficiency gap. Better process control also improves factory yield and time-to-yield during a startup. A whole industry, including the multibillion-dollar KLA-Tencor, grew up around the need to control and improve yields in the semiconductor industry. Solar needs its own KLA-Tencor. Metrology companies usually enjoy good margins and do not require lots of capital; the right company will deliver a good return on investment.

Another area to look at is enabling equipment. Throughout history, new equipment has been the source of improved efficiency and yield. Semiconductor technologies, such as ion implantation and Rapid Thermal Processing (RTP), deliver superior process control that could lead to improved efficiency and yield in solar. The challenge is throughput; the equipment company that unlocks the secret to higher speed tools could expect widespread adoption. Equipment is a high-margin business and could be a very attractive investment.

4. Find technologies that reduce capital intensity: As solar approaches the gigawatt era, the need for capital to build new factories could become a bottleneck. There are two methods to achieve lower capital intensity.

Process simplification is the most elegant option. A simple six-step manufacturing process typically requires half the capex of a more complex, 12-step process. Think about it: First Solar has low efficiency relative to c-Si, but is still the low-cost leader. The reason is process simplicity. A company that develops a simpler manufacturing process may be a great investment opportunity.

The brute force approach to lower capex is high equipment throughput. This opportunity applies everywhere in solar: frontend, back-end, silicon, and thin film equipment. If a system has twice the output, only half as many systems are needed. As the semiconductor industry has demonstrated, advanced manufacturing with high speed automated tools can increase factory output by 100% with as little as a 20% increase in cost. At this rate, each doubling of throughput results in a 40% reduction in capital intensity. Awesome! This equipment scaling benefit has been taken advantage of in other industries for decades but is still a great investment opportunity in solar.

5. Develop the "solar appliance": Designed for the residential market, this is a small (2kW to 10kW), self-contained system that cuts out the middlemen and includes everything needed for a simple do-it-yourself installation (Chapter 14). If someone figures out how to overcome the considerable obstacles, this could be a disruptive investment opportunity.

6. Develop technologies that reduce BOS costs: Since BOS costs now exceed panel costs, this is an important area of interest. No specific technology is recommended, but some ideas were discussed in Chapter 14. If a new product comes along that simplifies installation, reducing BOS mechanical and electrical costs, it could be a very good investment.

7. Develop technologies that reduce soft costs: If soft costs in the US were to fall to German levels, demand could rise to more than 100GW (compared to 3GW in 2012). The solar appliance (see #5) would be a huge step in this direction. In the shorter term, business model innovations such as third-party leasing can also bring down soft costs. In fact, the SolarCity lookalike that solves the commercial owner-occupier dilemma described in Chapter 15 could be the next company to hit it big. This is a huge Solar 2.0 opportunity.

8. Develop technologies that improve solar yield: As the industry moves away from the installed $/watt metric and toward LCOE, solar yield (the yearly power output of each installed watt) is becoming increasingly important.

Micro-inverters are one way to improve solar yield. Although Enphase has been doing fairly well, there is still a big opportunity for a cheaper, more reliable product. With an inverter on each panel, solar yield can improve 20% to 30%.

Trackers are panel mounting mechanisms that continuously tilt the panel to directly face the sun as it moves across the sky. Even though trackers can increase solar yield by as much as 15% to 25%, their reliability is still low enough and their costs high enough to impede adoption. If a cheap, reliable tracker, or a passive tracking technology were developed that eliminated moving parts (simultaneously reducing cost and improving reliability), it could dominate the market. Trackers could make a comeback in Solar 2.0 because of their positive impact on solar yield.

9. Longshot #1 - Find the "black swan": These last two ideas are a little farther out. The first is the elusive "black swan" of solar: very high efficiency at very low cost. Hundreds of millions of dollars have already been spent chasing these black swans. Just because one hasn't been found doesn't mean they don't exist. Alta Devices, for example, used GaAs thin film on a flexible substrate. The company got the high-efficiency part right (28.8% record), but the device cost 10 times too much. Other ideas, such as multiple junctions and tandem structures, have had the same problem. Alta was the pioneer; the kind of pioneer hit with arrows. It's much better to be the settler. The successful settler would start from the opposite direction—getting the low cost first, and then focusing on efficiency. That is what Wakonda tried to do, with thin film GaAs on a cheap roll-to-roll substrate. Unfortunately, it was underfunded and didn't make it. Still, if someone breaks the code on high efficiency with low cost, it is a grand slam opportunity.

10. Longshot #2 - Invent no-glass, no-encapsulant panels: Materials make up most of the manufacturing costs, and silicon, glass and encapsulants are the dominant materials. There is little chance for additional volume cost reduction as these materials are made in huge volume already. Eliminating them altogether, though, could bring costs way down, perhaps to as little as $0.25/watt. This would be transformative, but is it even possible? By building on existing thin film technology, it might be. Silicon is not needed in thin film, so that solves that problem. To eliminate glass and encapsulants, start with a roll-to-roll production line, then add two additional steps: a polymer deposition (replacing the encapsulant), and a silicon dioxide/silicon nitride deposition (replacing the glass). This final layer can be designed to not only act as a hermetic seal but to serve as an anti-reflective coating, trapping in more sunlight. These deposition tools already exist, so the idea may be feasible.

The resultant material cost would be about half of what it is today, and the final product would be light and flexible. There are major questions about the feasibility, durability, and reliability of such a structure, but making it happen would be really exciting.

When is a Good Time to Invest?

Timing is everything, and now is a good time to invest in solar. Maybe that is a bold statement, and I'm a solar CEO, so perhaps it is a bit self-serving. This recommendation though, comes not from bias but from the data. Four elements determine the best timing to invest; solar is a hit on the first three, and improving its position on the fourth.

1. Business cycle: It sounds simple: Invest at the right point in the cycle. Unfortunately, investors following (and perhaps chasing) investment trends often get the timing wrong. The cycle in Figure 16.3 was discussed in detail in Chapter 9. Now let's look at how it can inform investment decisions. On the downstream side, it is obviously a good time to invest when the price is softening. Solar panels are less expensive and demand goes up. It is not a good time to invest in the upstream side because the low prices cause manufacturers to lose money or even go out of business.

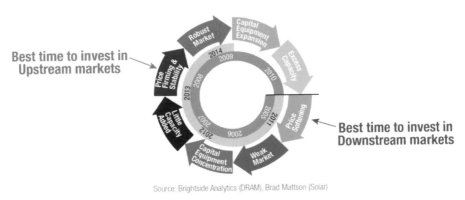

Source: Brightside Analytics (DRAM), Brad Mattson (Solar)

Figure 16.3 – When to invest in the business cycle

The time to invest upstream is on the opposite side of the cycle, when prices have started firming up. This is the signal for the next round of factory expansion. In 2013, for the first time in years, the price of solar panels went up, indicating that we are in the "price firming and stability" phase. That is the last stage before "robust market," and thus the best time to invest in upstream markets.

Unfortunately, as logical as this seems, it goes against "momentum investing." Many investors wait until there is significant movement in a market before diving in. This happened in solar. With the robust solar market of 2008, investors piled in on the upstream side...but it was too late. With the excess capacity in 2010 there was not enough time to recoup the investment before the inevitable decline. The profitable investors got in early (and got out in time).

2. Long-term market demand: Even though the profitability of some solar segments is weak, there has been 50% growth for a decade, and there is room for much, much more: Solar is still not even close to accounting for one percent of global electricity. In fact, the basis of this book is that the low prices that are causing angst for some suppliers are actually driving this incredible demand. So if we recognize that low prices have hurt profitability, we must also recognize the upside: Low prices have created high demand.

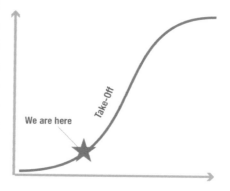

Developing industries typically travel the classic "S" curve; they start out small, then grow at high rates for years or decades before flattening out to grow at close to the world's average GDP. I experienced the S curve in semiconductors, and it was a wild ride; almost 30 years of 20% growth. We are just at the beginning of a 30- to 40-year growth phase of solar. That makes the present a good time to invest.

3. Availability of IP and talent: In the heyday of solar investing, around 2008, the deal flow was great. It seemed like a great time to invest; in fact, it wasn't. Valuations were inflated, quality deals were few and far between, and inexperienced management teams wasted a fortune. The inevitable correction was ugly, as shown in the drastic drop-off in Figure 16.4.

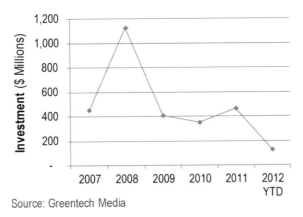

Source: Greentech Media

Figure 16.4 – VC investment in solar

In recent years there has been underinvestment—and this creates opportunities. Because of the overinvestment from 2007-2010, an unprecedented amount of technology is available for pennies on the dollar (some even for free). The IP, patents, know-how, equipment, facilities, and even the most important asset – people – are all available. Experienced investors will tell you that good investing is mostly about picking the right people. Others will say it's *only* about picking the right people. In any case, as a result of consolidation in the industry, some of the best solar executives in the world are available. This is a buyer's market for world-class management talent.

4. Follow-on capital: This is a challenge. There are capital-light segments within solar, and the best investments reduce capital intensity, but a lot of capital is still required. Initial investors need to believe that in the later stages, when large amounts of capital are needed, the money will be there.

It has become common knowledge that the VC business model is not the best fit for cleantech in general, and this is true for solar as well. VCs may not have enough capital to carry the investment through to a profitable exit (via IPO or acquisition). This doesn't mean there isn't a workable investment model. VCs need a better hand-off strategy, such as the one that has developed in pharmaceuticals. In the pharmaceutical industry, it is known that an IPO exit might be unlikely because the regulatory environment can extend product development well beyond typical VC investment timeframes. VCs and Big Pharma companies needed to create alliances to affect an orderly hand-off. Big Pharma made it clear what startups needed to do in order to qualify for the transition; the startups and VCs then drove in that direction.

Solar and many cleantech industries have not yet developed this hand-off process. It is not too late. It is happening now, less effectively, through M&As such as the SunPower-Total relationship and the GE-First Solar relationship. This is the counterpart to what Big Pharma did to create access to a large pool of capital; Big Energy is stepping in and helping finance the growth of solar. Many energy companies, along with firms in the materials, equipment, semiconductor, and electronics industries, believe solar has a promising future, and they have billions to spend.

Another reason more capital may flow into solar is the "greed vs. fear" dynamic. When there is bad news, fear dominates and investors flee. When good news is flowing, investors start seeing dollar signs. In the case of solar, there is a lot of good news. All the elements are in place for the follow-on capital situation to improve.

- Overcapacity is almost gone. As of early 2014, supply and demand are coming into balance with high demand absorbing excess capacity. In fact, most Tier One manufacturers have backlog and must add capacity.

- The perception of solar is improving. Even Warren Buffet is investing in solar, and he carries weight in the investment community. High growth rates certainly help.

- The necessary rationalization of the market has occurred, with weaker companies exiting and survivors prospering. Profits are back.

- There are big solar successes and exits. SolarCity had a great IPO, and as of Q1 2014 was up more than 600% from the IPO price. Enphase, another IPO, is also doing well. The stock market has picked up, but solar stocks have way outperformed the market. *In fact, the top ten solar stocks in 2013 were on average up over 300%.*[47] How can analysts ignore those kinds of returns?

Risk Mitigation

Before moving on to "Who Should Invest Where," a few words about risk. This section may be as much for entrepreneurs as for investors, but risk mitigation is important for everyone.

First off, successful entrepreneurs are not necessarily risk takers. In fact, the best ones go to great lengths avoid risk when possible, and to minimize those that are inevitable. Investors should look for management teams that understand this.

Next, scaling itself is not too risky, but premature scaling is very risky. Contrary to popular belief, it is usually not expensive to develop technology, including solar technology. It *is* expensive to scale it. Scaling is not, however, required to de-risk the technology. Both of solar's key performance metrics—high efficiency and low cost—can be proven at low scale and low burn rate (run away from any CEO who says otherwise).

The major cost reductions in scaling, like volume discounts, are easy to calculate and factor into a cost model. It is not necessary to scale up to prove volume discounts. The important manufacturing costs are built into the solar cell architecture and equipment design; these risks can be evaluated at low cost.

The lesson of recent history—and this cannot emphasized enough—is that poor risk management has been misinterpreted as excessive capital intensity. The solar industry is capital-intensive, but to nowhere near the level that investors perceive. Investors should look for managers who embrace the value of strong risk gates, and who are data-driven.

Finally, scaling is just one form of execution risk. There are four types of risk to manage: market, technical, financial, and execution, and ways to minimize each.

Market risk: Markets cannot be managed or controlled, so it is important to select or segment the market to avoid risks. It's best to avoid the following:

- A competitor that is a 900-lb. gorilla; you can get stomped too easily.

- A market that is "emerging," meaning it may never develop; don't try to develop it!

- Areas with little chance for differentiation. Customers have little incentive to switch to a new player, and there is little profit potential if they do. Low differentiation leads to intense cost competition and low profits.

- The product or service requires the customer to change behaviors. This rarely works, because people avoid change.

Of course it can be difficult to avoid all four conditions—big, exciting markets tend to attract competitors. Another way to mitigate the market risk is to enhance the product offering, making it more competitive. This means setting the bar higher on technical goals for that product. For example, do not target to be only 20% better than the competitor. Require your team to produce an 80% advantage. This reduces market risk, while increasing the product differentiation and future profitability.

Technical risk is mostly under management control. The simplest way to mitigate technical risk is by developing alternative pathways. This may be heresy to investors looking for entrepreneurs who have an intense focus on just one thing. But I know of almost no highly successful company that started and finished on the exact same pathway. Intel started out in DRAMs before switching to microprocessors. Cisco crashed several times. Neither of my own startups, Novellus nor Mattson, would have reached the IPO stage without major course corrections. The best way to mitigate technical risk is to have multiple shots on goal. Sure, there's a little extra R&D expense involved, but it's trivial compared to the risk mitigation value. My favorite saying in entrepreneurship is Darwin's quote: "It is not the strongest of the species that survives, nor the most intelligent that survives. It is the one that is the most adaptable to change." One can adapt to change by having options.

Financial risk, like market risk, is difficult to control, since you cannot force someone to write a check. But de-risking at each stage of development (see below) is a first step. Second, CEOs need to be prepared to devote half of their time to fundraising. Look for at least three independent sources of financing, such as venture capital firms, strategic investors (companies), overseas investors, and government sources. This is risk minimization through diversification; it is unlikely that all these capital markets will dry up at the same time.

Execution risk, including the scaling risk already discussed, is highly correlated with the quality of management. The mitigation strategy is simple: Hire good management. That is, of course, easier said than done. In a startup situation, do not hire inexperienced people. Startups are not the place for newbies, trainees, or postdocs. Get experienced people who have done it before, and pay for the best. Buy the learning curve; don't relearn it.

Who Should Invest Where? – *Stages of Investment*
The 10 profiles covered the "what" to invest in, but the "who" is also an issue; this will vary depending on the stage of development of the company. Consider the typical development cycle shown in Figure 16.5. Different investors are appropriate for different stages, based on their capital resources and risk appetite.

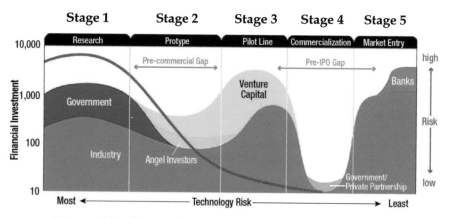

Figure 16.5 – Stages of investment - funding gap - risk profile

The solar industry has managed risk poorly. The red bar in the graph is a suggested level of "appropriate" risk at each stage of development. The biggest challenge has been to make sure most of the risk is eliminated before moving from the inexpensive prototype, or concept and feasibility (C&F) stage, into the much more expensive pilot line stage. On this basis, let's look at each stage, the level of risk, and the right investors.

Stage 1 – Basic research: New technologies start with basic research. This stage is highly risky, as most research does not result in a commercial product. At this early stage, the government, through the national labs and research universities, may play an important role. The bulk of basic research, though, is carried out in industry by huge, multinational firms. This stage is not interesting to most investors.

Stage 2 – Concept and feasibility: The C&F (or prototype) stage answers the question of whether a technology can solve a relevant industry problem. It is the most crucial of all stages, but often bypassed too quickly. This is where the critical risks, particularly the economic ones, should be identified, tested, and eliminated. The C&F gate must be a strong one: Don't let anything get past this stage unless the risks have been addressed. Looking at the red line in Figure 16.5, the correct objective in C&F is that 80% of the risk is eliminated before going on to the next stage. In the mad rush to "first revenue," and with a huge influx of VC money, many solar companies skipped risk management — and investors lost billions. C&F is not expensive, so this low-budget, highly leveraged activity is the perfect stage for venture capital investment.

Stage 3 - Pilot line: Here the focus moves from technical risk to scaling risk. This phase is much more expensive than C&F. It's important to minimize those expenses by picking the right scale. High enough scale to address key scaling risks (proving scaling works), but as low as possible to minimize expense. Stage 3 often involves unforeseen delays, so expense control is important.

In the pilot line stage, the capital costs start increasing. This is a good time to include strategic investors, with their industry knowledge and deep pockets, rather than rely solely on VCs.

Stages 4 and 5 - Commercialization and market entry: These are the final and most expensive stages. In fact, many companies will actually enter the market near the tail end of the pilot line stage in order to generate revenue as soon as possible. It is important to acknowledge that marketing and sales expenses are like R&D; there will be a lot of spending before there are a lot of results...way before. As a result, operating expenses always ramp up long before revenue does. Therefore, a lot more capital is required in this stage; strategic investors, private equity, and even the public markets should all be considered to assure adequate financing.

The Funding Gap

There are investors available for all the early stages of investment, but there is a problem in the Stage 4 commercialization phase. VCs have exited this area at the same time that strategics have taken a "let's-wait-and-see" approach. It is considered a little early for either private equity investors or an IPO. As a result, there is a considerable gap in investment, called the pre-IPO funding gap. This funding gap starts a negative chain reaction, as even earlier Stage 2 (VC) investment can fall off because they don't see a continuous investment pathway to the market. This pre-IPO funding gap has to be filled to have a healthy investment pipeline. Here are four ideas to fill it:

1. *Lower the bar*: Sure, capital is required, but not at the magnitude many assume; it is a myth that it takes $400 million for commercialization. Many high-profile failures, such as Solyndra, Nanosolar, and Abound, did not fail because of scaling issues. They failed because the technologies either didn't work or were too expensive. This should have been caught in the C&F stage. These were problems of risk management, not capital intensity. The company I am now working with only needs $100M to go through this commercialization stage. A lot, but it is a manageable amount.

2. *Install gates*: In order to "lower the bar," reducing funding requirements from $400M to $100M, discipline is required among entrepreneurs, managers, and board members. Strong C&F gates are needed to prevent premature scaling efforts which are so costly. Deal with *all* the risk first, scale-up later. Gates are crucial to minimize funding requirements.

3. *Find and nurture Stage 4 investors*: The investors that have left Stage 4 need to be enticed to come back, or we need new ones. Without their money the investment pipeline becomes clogged, and the innovation that fuels industry growth is stunted. There are several channels to explore to cultivate Stage 4 investors:

Large VC or growth funds are an outgrowth of the cleantech VC phase, and were created just to take cleantech companies from Stage 3 to Stage 4.

Private equity (PE) is not normally associated with early stage, pre-revenue companies. Still, the size of investment required and the size of these funds is a good match, if a company is appropriately de-risked.

Family offices are the investment funds created by wealthy individuals or families. These funds tend to have a longer time horizon and some are truly committed to having a positive social and/or environmental impact, making them a good fit for solar. Like PE, the size of some of these funds is a good match for Stage 4 investing.

Strategics are large, multinational corporations that have the money, knowledge, patience, channels, and strategic reasons to get involved. These are the most likely candidates for Stage 4 investment. Appropriate strategic companies can be found everywhere in the solar value chain: materials, equipment, EPC, service, finance, and energy companies, but also in associated industries like semiconductors, flat-panel displays, and the electronics industry.

Governments often take the lead in building national infrastructure, and they have supported solar in every stage of development. In the US, the loan-guarantee program has been a major assistance to Stage 4 companies. In fact, although the unsuccessful ones got all the bad press, the program has been overwhelmingly successful. The US loan program was and is still a very good idea to bridge the pre-IPO funding gap. Unfortunately, the best place to go for companies seeking government investment is not the US, but China, Russia, India, or Saudi Arabia. These governments view solar as a technology of the future.

4. **Perfect the hand-off:** There are not enough mechanisms to effectively involve Stage 4 investors in the hand-off from VCs. VantagePoint Capital Partners (VPCP) understood this and involved strategics very early in the deal flow. They brought them in as limited partners, had joint sessions to review deal

flow, shared investment roadmaps, and even hired "ex-strategics" as venture partners. This is a good template for success. The industry needs more of this approach. Treat the process as a workflow. Bring the strategics in early, and inform them all along the way. Companies need to learn effective ways to work with strategics.

VCs, entrepreneurs, and boards of directors need to better prepare to bridge the pre-IPO gap, taking companies through the proper de-risking process and involving strategics in those decisions. The pharmaceutical industry has been successful with this, and solar could benefit from this approach as well.

~ ~ ~ ~ ~ ~ ~ ~ ~ ~ ~

There is one category of investors that hasn't been discussed. When government and markets break down, there is an opportunity for individuals to step up. Solar energy can lead to the democratization of energy and may benefit everyone on the planet. Supporting this effort is not a matter of philanthropy. It should be extremely profitable so there is a chance here to accomplish the classic "doing well while doing good."

To: *Billionaires who care*
Subject: *Help get Solar 2.0 rolling*

Together, with technology already available, we can begin to halt global warming, save countless lives by reducing pollution, democratize energy, and leave the world a better place for our children.

*I truly believe that in this case **one person can make a difference**. That person could be an industry leader like Elon Musk, a financial leader like Vinod Khosla, a political leader like Bill Clinton, or it could be Oprah.*

It just takes one visionary with the desire and courage to do it.

Capital, both financial and human, is the life-blood of any industry. If we do not ensure that capital flows into solar, the industry cannot prosper. There is no shortage of investment opportunities.

So let's start the conversation. We need you!

Sincerely,
Brad Mattson

Chapter 17
Policy Roadmap

The world is a laboratory for solar policy. There are dozens of countries with solar policies, and over a decade of results. The findings are unambiguous. To kick-start a successful solar industry, use a feed-in-tariff (FIT) to stimulate both demand and supply. For the US, which has both tremendous energy resources and very high consumption, an aggressive energy policy (which includes transportation) makes sense. Although US reliance on Gulf oil is declining, it is still substantial[48] – around 20% of all imports. A forward-thinking and imaginative energy policy could pay for itself, both in economic and political terms. This chapter describes the different kinds of solar policies, and then discusses what different countries do. Finally, it makes specific recommendations for the US.

Governments have played a pivotal role in the emergence of solar, but how they have done so has varied widely. Four groups of policy tools have been used: standards, incentives, financial assistance, and market-based solutions. Let's look at how each policy works.

Standards: Renewable Portfolio Standards (RPS), also known as Renewable Electricity Standards (RES) or Renewable Obligations, set targets for renewable energy production in a specific region. They can be voluntary or mandatory. In the mandatory case, utility companies must produce a certain percentage of renewable energy, and are penalized if they fall short. In effect, this is a stick, not a carrot, which prods utilities to increase the supply of renewables. RPSs are in force in many countries, including Britain, China, Italy, Poland, Sweden, Chile, and in about 30 out of the 50 US states.

RPSs are often accompanied by Renewable Energy Certificates (RECs), which were designed to create a larger and more efficient market for renewables. The EPA defines RECs as the property rights to the environmental, social, and other non-power qualities of renewable electricity generation. An REC, and its associated attributes and benefits, can be sold separately from the physical electricity associated with a renewable-based generation source.[49]

RECs provide a method for utility companies to comply with RPSs without installing renewable energy themselves. RECs can be created by a developer in a great solar area like Arizona, then bought by a utility to meet their RPS mandate in a sunlight-challenged place like Seattle. This system helps ensure lower-cost production of renewable energy. RECs can also be purchased by corporations or anyone interested in supporting renewable energy. Many environmentally conscious corporations like Google and Intel have purchased significant numbers of RECs in order to move toward carbon neutrality. On the selling side, RECs provide an additional income stream, and therefore a further incentive for developers to build renewable energy systems.

Incentives: These are the carrot as opposed to the RPS stick. Incentives come in two major forms: tax incentives and feed-in-tariffs.

Tax incentives offer producers or consumers tax credits or exemption from certain taxes based on the operation or installation of solar power systems. The use of the tax incentive can be complex, and leads to interesting structures like REITs and MLPs (see Chapter 15). New ways to use tax credits are likely to emerge, but the credits themselves are straightforward.

Feed-in-tariffs (FITs) are the most important incentives globally. FITs led to the rapid expansion of solar in Germany, Spain, Italy, and other countries in Europe. In 2012, Japan joined the FIT club. A FIT guarantees owners of solar power systems a known return on their investment for the life of the contract (up to 25 or 30 years). Governments vary the FIT incentive from time to time to manage the amount of solar power coming on stream. It will often start with a relatively high rate of return to encourage the build-up of solar capacity. In later years, the FIT is reduced as targets are met. Also, as solar installation costs decline, fewer incentives are needed to motivate installers.

This is an important feature of a well-designed FIT; a too-high FIT, or one that is not adjusted as required, can lead to economic problems. That has been the case in Spain which, after an aggressive start, cut payments drastically in 2012, basically reneging on payment promises and leaving solar investors stranded. The once-booming Spanish solar industry is now suffering.[50] Spain's painful experience is a clear case where a FIT went wrong, but these tariffs have succeeded in many other countries. The lesson is to design the FIT properly: set the right initial return and reduce it once demand meets the goal.

Countries that have implemented FITs, even poorly, have seen tremendous solar expansion. Here are some of the benefits of an effective FIT:[51]

- It is cost-effective: In 2008 Germany's additional cost for its national FIT was 3.2 billion euros. The German Federal Ministry for the Environment calculated the return due to the FIT as 7.8 billion euros from reduced fossil and nuclear fuels purchases, and 9.2 billion euros saved from the avoidance of external costs.

- It does not impact budgets or taxes: A FIT is not a subsidy, and no public debt is needed to fund it, making it a self-sustaining mechanism.

- It is simple: The policy requires utilities to buy solar power at a certain rate, so it is easy to administer. There is minimal bureaucracy and red tape.

- It reduces solar's total installed cost: Economies of scale coming from increased volume of business leads to reduced installation costs. Also, certainty of returns leads to risk reduction, and therefore lower cost of capital.

- It creates jobs and encourages private investment.

In fact, FITs work so well that the key issue is slowing them down at the appropriate point. It will be interesting to see how things work out in Japan, which has recently implemented an aggressive FIT in the wake of the 2011 tsunami and nuclear accident in Fukushima. The country should be able to benefit from the experiences of other countries; in fact, in 2013, solar installations in Japan were second only to China.

Financial incentives come in two forms. First, governments that want strategic industries to explore new technology commonly offer R&D support. This occurred in lasers, semiconductors, the Internet, and now solar. These grants may fund 100% of a project or they may be in the form of matching funds to ensure that the recipient is committed and confident. Second, governments can offer incentives designed to lessen the financial burden of capital-intensive projects, such as solar factories or power plants. These are endeavors that, even if economical in the long term, may be difficult to finance in the short term. Here are four common financial incentive programs:

> *Loan guarantees*: These essentially fill in the Stage 4 financing gap (Chapter 16). The guarantee requires a third-party bank to provide the actual loan, with the government guaranteeing against a default. The strategy is to improve the probability of a successful loan by employing banking industry expertise. The US and China, in particular, have successfully used this option.

Net metering: Net metering incentivizes end users to install solar power systems. It basically allows users' electricity meter to run backwards, crediting users when their system produces more electricity than they are using, thus lowering their electricity bill. This policy is usually limited, in that under no circumstance does the utility pay the user. In other words, the electricity bill can only go to zero. So, unlike a FIT, there is no incentive to produce more electricity than can be used on site. Still, net metering encourages individuals to install solar power as a way to reduce or eliminate their electricity bill.

Property Assessed Clean Energy: PACE allows real estate owners to borrow money from municipalities to install solar power systems and then pay that back over 15-20 years via an increased property tax bill. Municipalities can finance PACE through bond measures. The risk to the bond owner and municipality is low because of the fixed return embedded in the property value. PACE is an exciting policy tool, because it matches the long-term value of solar to another long-term value—the building it is on. In 2008 Berkeley, California became the municipal pioneer of PACE, and since then 30 states have passed PACE-enabling legislation. In 2010 PACE was dealt a blow when Fannie Mae and Freddie Mac refused to back mortgages with PACE liens on them. While solar advocates address this setback, it should be noted the impact is mainly on residential properties. On the commercial property, there are 25 active commercial PACE programs in 10 states.

Community solar allows aggregation of demand to lower the cost of solar power and make it more accessible to millions of homeowners (Chapter 15).[52] This has not seen widespread adoption yet. Because this requires the involvement of utilities, community solar is being pioneered mostly in municipalities where the city owns the utility. There are over 30 community solar projects in 12 different states, with success stories in communities like: Ellensburg, Washington; Sacramento, California; and Tucson, Arizona.

Market-based solutions: These policies use the efficiencies of markets, in particular the price signal, to properly value assets so that purely economic decisions drive adoption of the appropriate technologies. The techniques vary, but the common element is to assign either negative or positive economic value (i.e., cost or benefit), to something that is currently not valued, but should be included in the economic analysis. On the "cost" side, for example, one could price in the environmental cost of greenhouse-gas emissions or the healthcare costs of pollution, via a carbon tax. On the "benefit" side, one could price in the positive value of solar, by giving extra credit or value to solar energy.

The two most prevalent concepts are *cap-and-trade* and a *carbon tax.* In a cap-and-trade plan the regulator sets a "cap," or limit on the total amount of CO_2 that can be emitted each year. This cap is divided into permits or allowances that are available for sale. Firms are required to hold permits equal to their CO_2 emissions. Firms with too-high emissions can buy permits from those who have permits to spare. This allows the most cost-effective solutions to flourish and rewards the most carbon-efficient firms.[53] To reduce overall emissions, the cap goes down year by year, allowing companies to plan their emission-reduction strategy.

Cap-and-trade systems can work; the US has used this mechanism since the early 1990s for sulphur dioxide, and it worked even better than its creators thought it would.[54] On the other hand, Europe's carbon cap-and-trade system has not worked well, for a variety of technical reasons. Cap-and-trade does not automatically lead to increased renewable energy generation, and requires extra government intervention and management. The devil is very much in the details.

A carbon tax, by contrast, is relatively simple and can be implemented quickly. Carbon emissions are taxed to price the environmental and social cost of introducing CO_2 into the atmosphere. Emitters can reduce their tax liability by switching to low-carbon technologies or by capturing, or "sequestering" the carbon. Businesses pass along their higher costs, if any, to consumers through higher energy prices, but this can be offset by reductions in other taxes. It is important to make the carbon tax net neutral to consumers, so that its only impact is to shift profits from carbon producers to carbon eliminators.

The US already has carbon taxes of sorts on transportation and heating fuels. A rational tax policy could extend these to the use of fossil fuels for power generation. That would encourage the transition to non-carbon power sources. Here are some other benefits of a carbon tax, as opposed to cap-and-trade:[55]

- Carbon taxes lead to more predictable energy prices, whereas cap-and-trade may create price volatility, discouraging investments in low-carbon technologies.
- Carbon taxes are transparent and easily understandable.
- Carbon taxes are less prone to manipulation by special-interest groups.
- Carbon taxes address all emissions of carbon, where cap-and-trade tends to target electricity generation, which accounts for less than 40% of emissions.
- Carbon tax revenues can be returned directly to the public.

Both cap-and-trade and carbon tax are examples of *cost-based* policies. For a *benefit-based* policy, one idea is the "value of solar tariff" (VOST). A utility in Texas, Austin Energy, has implemented a VOST that pays solar owners for the benefits their distributed solar system provides compared to a centralized, carbon-emitting generator. These compensated benefits include: avoided transmission and distribution costs, the value of enhanced grid security, environmental and health benefits, and other social and economic factors. Providers of solar energy get a VOST payment from the utility for the power the solar system puts in the grid. This policy is still new, but based on Austin Energy's work, the credit could be between $0.11 to $0.13/kWh.[56] VOST received a very significant boost when a court in Minnesota ruled that, based on the VOST value, the local utilities should use solar energy over natural gas when adding to the energy supply.[57] This set a precedent for valuing the benefits of solar energy.

Now that we have reviewed the menu of the policy tools available, the next section looks at what is actually used around the world.

Global Policies

National policies vary broadly, and they can be difficult to track. For example, the PV-Tech Tariff Watch[58] lists 42 countries with FIT programs while PV magazine shows 56.[59] Ren21, the Renewable Energy Policy Network for the 21st Century, maintains a reasonably up-to-date compilation of global renewable energy policies, including an interactive map.[60] A few observations from that data:

- Higher-income countries maintain a number of incentives, including FITs, direct grants-subsidies, and tax incentives, often used in combination.

- Lower-income countries rely more heavily on tax incentives, but there are a few FITs.

- Tradable renewable energy credits, renewable portfolio standards, and net metering are significantly less prevalent.

- The US and Australia are outliers—high-income countries that do not use FITs. Of the top 15 countries (ranked by cumulative solar installations in Figure 17.1), all of them, except the US and Australia, use FITs as their primary incentive.[61]

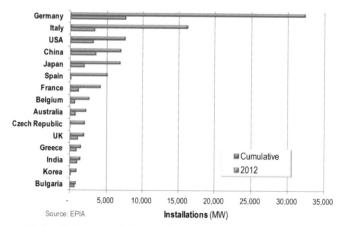

Source: EPIA

Figure 17.1 – Solar countries ranked by cumulative installations

Germany stands out with a well-designed FIT that started in 2000 and has been modified periodically to reflect lower installation costs. The FIT worked as intended, promoting a significant volume of installations, which led to volume-related cost reductions.

Other European countries copied this model, with Spain and Italy reaching grid parity much faster than expected. As mentioned earlier, Spain's FIT was poorly designed; the initial level was unsustainably high and there were no provisions to lower it. Small, ground-based projects could receive almost six times the average electricity price. As an entirely predictable result, supply surged to quickly overwhelm the budget, and an economically-stressed government slashed FIT rates and crashed the market (Fig. 17.2). Spain's experience shows both the positive and negative effects of FITs; they will dramatically increase installations, but the economics have to be managed to avoid over-heating.

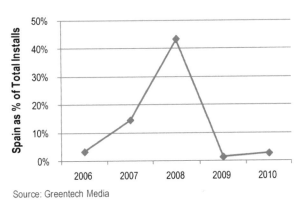

Source: Greentech Media

Figure 17.2 – Spain as % of global installs

US Policy

The US is a melting pot of policy, with a little of everything, varying by region, state, and even municipality.[62] There has been some local experimentation with FITs. As of late 2013, there were about six states and a few municipalities with FITs, but these are not working well because rates are too low to attract investment. The successful experience in other countries has not yet transferred here. US policy includes:

- An Investment Tax Credit (ITC), which provides a one-time credit of 30% on the cost of installing a qualified renewable energy system

- A Production Tax Credit (PTC), which is an ongoing income tax credit based on kWhs of energy produced each year

- State and local tax incentives, including sales and property tax benefits, local tax credits, and rebates

- RPS provisions in about 30 states, led by California, where utilities must provide 30% of electricity from renewables by 2020

- Grants from government agencies like the Department of Energy and Department of Defense to fund both basic research and to transfer technology into high-volume manufacturing

- Loans aimed at specific groups, such as low-income or veteran homeowners, local and state governments, and non-profits

- Loan guarantees for building solar factories and power plants, the most famous of which was the controversial "1705" Federal Loan Guarantee Program, which guaranteed $13.25 billion in loans to 16 solar projects.[63] As a whole, these loans were a huge success, with only 4% of the dollars invested resulting in failure. That success rate is a much better than VCs…by far.

- Net metering has been very successful in California, but has also become a battleground in many states, pitting solar advocates against the utilities. Utilities argue that homeowners can (if they install a big enough solar system and "run the meter backwards" during the day) end up paying nothing for electricity, although they are still connected to the grid and use grid electricity at night. Utilities want to be compensated for this service. It's a reasonable argument. Others argue the solar benefit to the grid outweighs the value of the storage function (consider the VOST data discussed above). Recently in Arizona, the utility companies won a decision to get payment for this grid service, but it was only $0.70/kW installed, or about $5 per month for a typical home. So both sides won…kind of.

Policy Roadmap for the United States

The roadmaps in this book have generally been applicable globally, but on the policy side this is difficult, as every country has different requirement. So specific policy recommendations will be made only for the US.

The US has a wide variety of policies, but no plan—and that lack of planning shows. Although the United States is #1 in electricity consumption globally, it ranked only 4th in solar installations in 2013, behind China, Japan, and Germany. US policy not only doesn't match its energy needs, it doesn't meet the country's moral responsibility, considering that the US is, cumulatively, the biggest carbon generator of all time.

What should the US be doing? I'll take a stab at answering this, but fair warning: In this section you will probably find something to like, something to consider, and something to hate. That's okay. We *need* to have these conversations and weigh differing points of view. The US can only reach the appropriate energy goals if it discusses and then implements fundamental changes in the way it makes and uses electricity. Here is a policy roadmap to get the conversation started.

1. Admit we need an industrial policy for energy: The United States has always distrusted the idea of a centralized industrial policy. However, it is naïve to think that we don't have one. It is just a haphazard one, often the result of decisions made decades ago, hidden in the tax code, or supported by entrenched interests. We need a comprehensive national policy with a clear goal such as: *The United States should have a renewable energy-driven economy (including transportation), that achieves permanent energy independence and both environmental and economic sustainability, becoming a net energy exporter by 2020.*

2. Form a SWOT team: The US has the natural, technical, business, and financial resources required; what we are missing is the strategy. One approach is to treat solar like any other strategic planning exercise, and do a SWOT (Strengths-Weaknesses-Opportunities and Threats) analysis. The US energy SWOT might look like this:

Strengths
Technology, good R&D
Innovation
Huge natural-gas reserves
Business infrastructure
Great wind/solar resources

Weakness
No coherent energy policy
Political gridlock
Huge appetite for energy
Weak in manufacturing
High-cost land and labor

Opportunities
Tesla, leading e-car company
US is leader in thin film solar
The US needs jobs and exports
New business models

Threats
China targets energy industry
Saudi Arabia commits $100B
Climate change causing havoc
Entrenched energy entities

Based on this SWOT analysis, here are the critical success factors for a new, improved energy policy:

3. Choose the right energy mix: The exact details of what the energy mix should be are not detailed here. The numbers are complicated and depend on how aggressive the policies are to stimulate renewables. Instead, here is a broad outline.

Start with implementing an "all-of-the-above" strategy; expand the supply of cleaner energy sources in order to reduce the use of coal. This includes adding nuclear and natural gas power plants as needed to cover the energy gap until solar and wind scale. Even with a FIT to stimulate renewables, nuclear is likely needed to fill the energy gap. The long-term goal should be to use natural gas for exports (as discussed later in this chapter). In the short term, though, natural gas is the cleanest conventional fuel, and therefore a great interim solution until renewables can meet demand.

4. Make energy efficiency a priority: "Negawatts," or conserving energy, has the highest return on investment of any option. There are great policies in place to learn from. In California, for example, per-capita energy consumption is 30% lower than the national average. In fact, it has almost been flat for about 30 years (Fig. 17.3), and the state has done well economically, recording the sixth-highest growth rate in the country since 2009.[64] California could and should be used as a template for a national energy-efficiency policy.

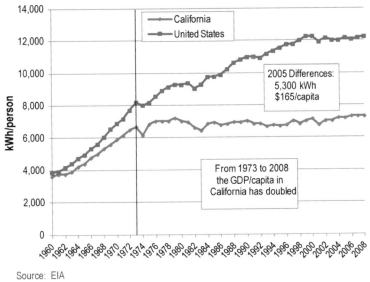

Source: EIA

Figure 17.3 – Electricity use per person US vs. California

5. Use a nationwide FIT to expand renewables and distributed generation: Start with the global best practice, meaning a national feed-in-tariff. This is central to a new, improved, energy policy; a FIT could develop wind and solar on the scale of tens of gigawatts a year, rather than the single digits now seen.

The goal of the FIT should be to also foster distributed generation (Chapter 3). This is done by having higher tariffs for smaller-scale, distributed generating power sources. Solar is especially amenable to DG. The larger goal, though, is to make renewables the cornerstone of future growth. While the US needs an all-of-the-above strategy, not all energy sources are equally valuable. The FIT should strategically drive investment towards those renewables which best align with needs and resource availability.

A FIT also has the benefit of being market driven. Market forces, driven by good ROI (return on investment), bring capital to the industry with little government intervention. Just set the right rate, and the market will take care of the rest. Germany's experience, as opposed to Spain's, shows how to do this properly.

6. Don't level the playing field; tilt it toward renewables: Believe it or not, the current playing field is tilted towards oil and gas. A recent report by DBL Investors[65] analyzed the historic levels of government support for various energy sources (Fig. 17.4).

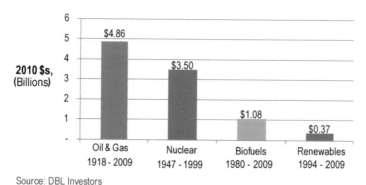

Source: DBL Investors

Figure 17.4 - Historical average of annual federal energy subsidies

As the chart shows, oil and gas have received annual subsidies significantly higher than any other energy source. Nuclear energy received the highest subsidies early in its development and adoption cycle. It can be argued that coal, oil, and gas industry tax benefits are no different from the tax benefits available to conventional manufacturing businesses.[66] However, the very fact that fossil-fuel tax provisions are permanently embedded in the tax code provides a stable market signal that renewable energy does not enjoy.

The non-partisan Energy Information Agency (EIA) provides a picture of the current status of incentives and subsidies in Figure 17.5.

In this, solar lags well behind oil and gas, coal, wind, biofuels, and energy efficiency in direct incentives and subsidies. In short, the playing field is definitely not level. The purpose of subsidies and incentives is to nudge people or industries in a direction that policymakers favor; that is why, for example, there is special treatment to promote saving for education and retirement. Surely it makes sense to incentivize new, strategically important areas, rather than old, well-established industries already at scale, with billions in revenue.

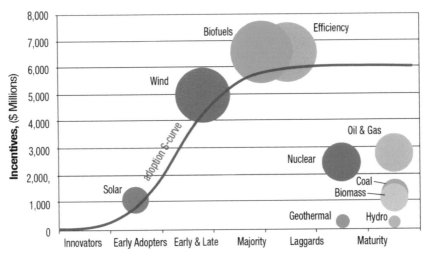

Source: USEIA (2011), Direct Federal Financial Interventions and Subsidies in Energy in Fiscal Year 2010

Figure 17.5 – Current US status of incentives and subsidies

It isn't good enough to level the playing field; *the playing field should be tilted towards renewables.* We have favored oil, gas, and nuclear for decades. It is time for renewables now!

US Energy Jobs - One specific suggestion to increase renewable subsidies is to reinvigorate and expand the successful 1705 loan guarantee program. It can be used not only for large, renewable power products, but for building domestic manufacturing capacity. Bridging the energy gap will require a lot of renewable energy products. Why not manufacture them here in the US, creating jobs for Americans?

7. Streamline adoption processes: Another way government can encourage adoption of renewables costs nothing at all; it is a matter of making the government easier to work with. In this picture, look at the paperwork required for a US rooftop installation, compared to Germany's two-page online application. No wonder that German installation costs are half those of the US.

The DOE has funded work to reduce soft costs, but a policy recommendation has yet to emerge. Adopting a simple national application and permitting package for qualified solar systems would be a great step. Building standards are another area where government could make things easier. As advocated by Solar Freedom Now,[67] a qualified standard rooftop system could include all UL-certified, pre-approved components, designed by pre-certified engineers and installed by licensed installation teams. Germany has proven that all of this is doable. We just have to copy their successful programs.

8. Export natural gas: Saudi Arabia has allocated $100 billion to develop solar, so that it can sell its oil rather than burn it. This is another example that the United States should follow. The long-term goal should be to export natural gas (see Chapter 2); domestic production is significantly more valuable in foreign markets than in the US. Of course, until wind and solar get to scale, natural gas will be needed to generate electricity, but this should be seen as a temporary phase. It is much more valuable as an export.

The Federal Energy Regulatory Commission should facilitate gas exports by accelerating approvals for the conversion of natural-gas import facilities to refinement for export.[68] Nine locations have applied for DOE authorization to export domestically produced natural gas and have yet to receive approval.

The EIA estimates that by 2020[69] the United States will be a net-exporter of energy, primarily oil and natural gas. This timetable should be accelerated. Minimizing the domestic use of oil and natural gas through the increased use of renewables will enable the US to export a larger fraction of its fossil-fuel production.

9. Create and pass the "Utility Restructuring Act": The utility industry, which is heavily regulated, is actively seeking to limit the connection of renewable sources to the grid. This is understandable; renewables do not fit their business model. But this means there is a mismatch between what is needed and what utilities are motivated to do. To change this dynamic, the utility structure needs to be changed, and the change needs to be dramatic. That is the idea behind the proposed "Utility Restructuring Act (URA)."

This URA does not yet exist, but my associate, Les Fritzemeier, drafted it to start the discussion. Personally, I think it's great. The Telecommunications Act of 1996 is the model; this deregulated the industry in fundamental ways, opening up markets to more competition, and leading to increased service options and reduced costs for users. The parallels between the communications and energy networks are striking. The Telecoms Act showed that entrenched systems could be changed to accommodate new technologies.

- The Telecoms Act allowed the FCC to preempt state or local legal requirements that acted as a barrier to entry for new technologies. The URA would give the DOE similar authority, to preempt local and state regulations that inhibit the growth and connection of renewable energy sources.

- The 1996 Act created obligations for incumbent carriers and new entrants to interconnect their networks. The policy established restrictions on incumbents who wanted to curb competitive entry by denying interconnection, or by setting terms, conditions, and rates that could undermine the ability of the new entrants to compete. The URA would create similar rules to ensure access to the grid.

- The 1996 Act required incumbent local carriers to make available, at cost-based wholesale rates, those elements of their network that new entrants needed. Under the URA, every qualified renewable-energy generator, regardless of scale, would be allowed access to the grid at cost-based rates. There is a fair cost associated with maintaining the transmission and distribution network; that cost should be recovered by the manager of the grid.

- The 1996 Act provided compensation for services rendered, requiring that inter-carrier compensation rates among competing carriers be based on the costs. A similar principle should be part of the URA, so that each link in the value chain makes a reasonable return. The policy must allow utilities and ISOs (Independent System Operators) to make money for base-load, storage, and distribution that the grid offers to renewable energy systems.

10. Incentivize the purchase of electric cars produced in the United States: An energy policy that doesn't address the elephant in the room makes no sense. To truly be energy independent we need to wean ourselves off foreign oil, which is used mostly for transportation. The best way to do that is convert our transportation energy source from oil to electricity. Tesla, Nissan's Leaf, and others have shown that it is both technically and economically possible. As renewable electricity generation expands, the country can start to reduce oil imports with the goal of eventually replacing oil altogether with clean, renewable energy.

Putting it all Together

Here is a summation of the US policy recommendations:

1. Develop a national energy policy.
2. Support an all-of-the-above energy policy, but emphasize renewables.
3. Make energy efficiency a top priority.
4. Institute a national FIT that drives distributed generation.
5. Tilt the playing field toward renewables.
6. Invest in renewable manufacturing to keep energy jobs in the United States.
7. Simplify permitting and develop national standards to reduce soft costs.
8. Export natural gas, using it sparingly and temporarily for electricity generation.
9. Pass the Utility Restructuring Act.
10. Move to electric cars, especially US-manufactured.

Of course whatever mix of policies we adopt, everything needs to be paid for. The combination of carbon tax, FITs, subsidies, etc. have to make economic sense. Some of these policies are revenue-neutral (no tax implication), but some will cost us. These costs have to be weighed against the cost of our energy dependence. Hundreds of billions of dollars each year are spent on energy imports (oil), and hundreds of billions more are spent defending energy regions. An unofficial estimate, developed when I was a VC at VantagePoint, is that the combined costs of export and defense were close to $800B/year. That is billions not millions. These numbers would have to be estimated with much more rigor, but when added to the cost equation, I'm confident the recommended policies will be a net positive.

~ ~ ~ ~ ~ ~ ~ ~ ~ ~ ~

Every one of these policy recommendations will likely provoke opposition. Change can be uncomfortable for people and organizations, and entrenched entities naturally want to protect their interests. That is understandable—and also irrelevant because we must make changes. Even if it is difficult, we need to initiate the discussion, and then take action. We need to protect the environment, and we need energy independence and energy security. The only way that will happen is for the American people to act—the subject of the next chapter.

Chapter 18
Public Support Roadmap

This chapter is for the American public: Your support is required if we want the US to become a leader in Solar 2.0. If you do not help, the future of solar energy will be controlled by others. From grassroots success stories we can learn how to find a common goal, the right messages, organization, leadership, committed people – and funding. This roadmap is a blueprint for how to galvanize support for the solar industry. It is a tall order, but the stakes are high. Let's do it.

To get the right policies, the solar industry needs public support from the 300 million Americans who are affected by the decisions of politicians, business leaders, and others in positions of power. Only "we the people" can unstick our warring political parties, and move them from their entrenched positions. Everyone needs to understand how the solar industry can help not just everyone here at home, but everyone on the planet. Voters need to become informed, make their views known, and take action. To facilitate this, the solar industry needs outreach, education, and a great public relations program.

Grassroots advocacy can drive significant policy change. It's been done before. Organized grassroots campaigns led to women getting the right to vote in 1920, to the civil rights reforms of the 1960s, and to massive increases in public awareness of and funding for breast cancer research in the 1980s. Think of MADD, Mothers Against Drunk Driving, which led major legal reforms and a significant change in the public perception of drinking and driving. In each case, someone got upset, took action, and was persistent. Over time, the facts and ethics became clear to the larger community. We should learn from their experiences. What do we need to do to succeed?

Set clear goals: Advocates need to visualize what success looks like, and then pull together in the same direction. An established "best practices" in the business world is to make goals S-M-A-R-T. That is, they need to be Specific-Measurable-Achievable-Resourced-Timed. For example, instead of setting "get solar adopted through distributed generation" as a goal, a S-M-A-R-T version would be something like "Establish policy to facilitate community solar to supply 10% of Americans energy needs by 2020."

Get the message straight: The communication process will involve a lot of politics, so even carefully crafted statements will likely be distorted by opponents. It is important that the message be clear and simple. It also needs to connect to people's lives, to the human issues that matter, and to motivate people to take action.

Connect to the heart and home: Facts matter, but emotion and values sometimes matter more. When looking at successful campaigns, it is clear that real energy comes from connecting with the public's values. Use positive emotions that appeal to people's better instincts, as Martin Luther King Jr., Nelson Mandela, and Mahatma Gandhi did. For a great example, just look to 2010, when the clean energy community recruited the trusted, health-focused American Lung Association to help defeat an anti-carbon policy ballot initiative in California. Using the tag line "Clean Air Saves Lives and Money," television advertisements featured mothers expressing concern about the business-as-usual approach to fossil fuels and their negative impact on air quality, and therefore children's health. This helped reframe the debate.

Solar advocates are typically passionate about their beliefs, but not everyone sees how solar issues connect to their family, their jobs, or their community. By showing how the issues hit home, the message can spread broadly, regardless of politics, region, or economic status. The focus on the solar industry's role in creating jobs is a great example of bringing the topic back home; the industry makes a case that matters in tough economic times. By closely tracking the growth of solar jobs and shedding a light on the industry's ability to create jobs, the solar community is making a case that matters.

Get the message out: Volume and repetition are important. State the case through as many media as possible: written, voice, video, and through multiple channels in each medium. Social media is particularly important. The lesson of contemporary politics is that knocking on doors must be accompanied by steady streams on Twitter, Tumblr, Reddit, blogs, and context-appropriate ad placements on major social platforms. It works.

People often discount the value of preaching to the choir. But activating the base is a tried and true tactic—and in fact, the base for supporting solar is large. A 2011 study by Kelton Research showed that 9 out of 10 Americans support the use and development of solar power, and 82 percent want more federal support for solar. And the support is broad—80 percent of Republicans, 90 percent of independents and 94 percent of Democrats agree that it is important for the United States to develop and use solar.[70] Also, the scientific community can make a big contribution by bringing forth relevant scientific facts and sharing the message in their circles. This is being done, but more volume is needed. As Penn State University climate scientist Michael Mann wrote in a 2014 op-ed in *The New York Times* entitled "If You See Something, Say Something," scientists are starting to take a more activist role, and that should be encouraged.

The message needs to be memorable. It is possible to stick to the facts, and still make them entertaining. With the exception of a fun tongue-in-cheek video series produced by SunRun that debunked the myth that solar was for hippies, and a clever NIVEA magazine advertisement in Brazil that contained a small solar panel that allowed beachgoers to charge mobile phones, it's hard to think of entertaining engagement around solar power. That should change.

Involve the public: Not just their hearts, but their wallets, hands and feet. In fundraising, the mantra is: "Don't forget the ask." Involving people in the solution is essential to getting a movement started. Don't just talk to people—involve them. Have a clear call to action, and propose different ways to get involved. Not everyone will march on the Capitol, but they can make a small donation, or write an email to their congressional representative (especially if that email is prepared for them). In early 2014, the Solar Energy Industries Association (SEIA) launched its "America Supports Solar" campaign. This encouraged people to "Shout Out for Solar," asking them to submit photos of themselves (or even their pets) with a message of solar support. Amplified through social media, the campaign increased SEIA's following considerably.

Get organized: One thousand individual efforts are much less effective than a coordinated effort of a thousand supporters. Advocacy takes a lot of work, people, and money, and it all needs to be organized. The problem is that there are too many support groups, mostly working independently of one another. A partial list of solar organizations includes: SEIA (Solar Energy Industries Association), AEE (Advanced Energy Economy), SEPA (Solar Electric Power Association), the Rocky Mountain Institute, the Clinton Climate Initiative, ACORE (American Council on Renewable Energy), Alliance for Sustainable Energy, Worldwatch Institute, Center for the New Energy Economy, ASES (American Solar Energy Society), InterSolar, the Solar Alliance, the Rahus Institute, AREI (American Renewable Energy Institute), CRS (Center for Resource Solutions), Bullitt Foundation, and EPI (Earth Policy Institute).

This list is just the tip of the iceberg. As one might suspect, duplication is rife. Promoting solar is a multifaceted effort that must be sustained over time; it is crucial to coordinate all of these efforts. It sounds silly to form a group to coordinate all of these groups, but maybe that's what we need. A bunch of passionate but uncoordinated efforts is unlikely to get the job done. The amount of human energy available is staggering, but it needs to be better directed.

Recruit a passionate, charismatic leader: Al Gore played an important role in elevating the problem of climate change with *"An Inconvenient Truth"* in 2006, but we did not have enough affordable solutions then. We have them now. We need a leader to emerge and point the way to the new energy future. Who will it be? It could be a billionaire like Bill Gates or Warren Buffet (both of whom already invest heavily in alternative energy), a celebrity like Leonardo DiCaprio, Bono or George Clooney, or a retired politician with cross-aisle appeal like Bill Clinton. I'm not sure if any of these candidates is a perfect fit, but someone is needed. A successful movement usually has a face attached.

Political leadership is also obviously important, and here the prospects are looking good. Solar energy is emerging as a "crossover" issue. Both libertarians and a new "Green Tea" splinter group from the Tea Party have been vocal in their support of the growth of distributed rooftop solar, and have opposed attempts by southern US utilities to increase rates on residential solar users. Continuing to build bridges with state and local politicians on the solar issue offers a great opportunity to not only garner support, but to blunt efforts that favor fossil fuels.

~ ~ ~ ~ ~ ~ ~ ~ ~ ~ ~

Public advocacy is important to build support for solar. Without that support, the policy roadmap doesn't have a chance. We have a good story to tell; let's tell it.

Chapter 19
The Solar Phoenix
"America Can!"

America can be a leader in Solar 2.0. The US is already succeeding in upstream and downstream solar markets, and can succeed in manufacturing as well by focusing on innovation and thin film technology. But will we? I think we can and will. We have risen to competitive challenges before, and we will do it again.

While most of this book has been global, the last few chapters have focused on how the US can succeed in Solar 2.0. Now, it is time to specifically address the topic of *"How America Can Rise from the Ashes of Solyndra to World Leadership in Solar 2.0."* Solar technology started in the US, and I'd like to see the US not lose yet another manufacturing industry to low-cost producers located elsewhere. This chapter will bring recommendations from all of the roadmaps together and select the best strategies for the United States, showing how the US can succeed, even in manufacturing.

The US should have plenty of motivation to play a key role in the solar industry. The transition to renewables is inevitable. And due to its leading role in the sector, solar technology will become an important strategic resource. Energy is critical for both economic growth and national security. Imagine if the United States were truly energy-independent. How many trillions of dollars would have been saved? How many soldiers' lives?

Based on current trends, though, the two future powerhouses of solar energy are likely to be China and Saudi Arabia. China has already spent more than $50 billion on solar, and Saudi Arabia has committed to spending $100 billion. Clearly, they want to be the major forces in this strategic technology, and have a plan for how to do it. The US does not have a goal or a policy. It needs both.

The Goal

What constitutes success? I think we should be ambitious. In his book *Built to Last,* Jim Collins argues that BHAGs (Big Hairy Audacious Goals) are what drive organizations to achieve incredible things. Here's a BHAG for the United States:

> *The United States should have a renewable energy-driven economy (including transportation), that achieves permanent energy independence and both environmental and economic sustainability, becoming a net energy exporter by 2020.*

Think about the positive results that might follow if the US met this goal: more job creation, less reliance on energy-rich countries, improved health, and a positive impact on both the US trade- and budget deficits. And these are just the obvious ones; there are always benefits from spin-off technologies, as we saw in the space program of the 60s.

Initially, the US might aim to accomplish what Germany and California are doing: 30% of electricity from renewable resources. To get there, the US would need to set goals for the domestic supply of wind, natural gas, and nuclear energy. Over time, the greater use of renewables would replace carbon-based energy sources, and release gas for export.

But *how* we develop our renewable energy is also relevant. America could import renewable energy products to reach our goal, but why replace importing foreign oil with importing foreign solar panels?

A central point of the strategy should include manufacturing. Success should be measured in part by the percentage of domestically manufactured solar panels. The benefit of domestic production is not just panel manufacturing jobs, but in jobs up and down the value chain. If manufacturing moves overseas, the supply chain follows. The solar industry already accounts for more than 100,000 jobs; there could be many, many more if the entire value chain were to remain in the US.

This is not to say that Americans should patriotically sacrifice by buying expensive or poor-quality American products. Far from it. The point is that with the correct policy and investments, and given America's strength in innovation, domestic production could be both higher-quality and lower-cost. Solar 2.0 is ripe to fulfill the "advanced manufacturing" initiative being promoted by the US government. China has invested $50 billion in older Solar 1.0 manufacturing technology. America can leapfrog those efforts with advanced Solar 2.0 manufacturing technology.

Domestic manufacturing also supports the net energy exporter strategy. When carbon-based exports like natural gas begin to wane, exporting renewable hardware like wind turbines and solar panels could help to maintain a positive energy balance of trade. So while domestic manufacturing is not strictly needed to achieve renewable goals, it is a good idea to create manufacturing jobs at home. If the US can do it without sacrificing quality or price, why not?

The US is a global leader in the business practices and market innovations that are needed in solar. Companies like SolarCity are likely to be emulated around the world. In fact, the US is already succeeding in much of the solar market. The downstream sector is inherently local; financing, labor, and other soft costs stay close to home. Of the BOS costs, only some of the BOS hardware — specifically, inverters — might be imported. Downstream businesses, then, are responsible for the bulk of US solar industry jobs, and new business models that reduce costs are important. The innovative culture of the US is a great asset, and because downstream is driven by innovation, America is well positioned to maintain strength in this sector.

Some upstream markets are prospering, too. The upstream business is basically about the supply chain: the materials, equipment, and components needed to install a solar power system. Manufacturers often prefer to buy components from local suppliers to keep shipping costs low, and because it is easier to keep an eye on quality control.

The US materials market is fairly strong. Everything needed to manufacture both silicon and thin film panels is available in the US. The country is a leader in polysilicon production, with suppliers such as MEMC (SunEdison) and Hemlock; industrial giants like Dow Chemical and DuPont are also committed to solar. Other basic materials, such as glass, encapsulant, aluminum, and steel are all made in the US, but often can be comparatively expensive. If the cost differential is higher than the savings from lower-cost local shipping, manufacturers may go overseas. These cost differences are dynamic and the end result remains to be seen, but a materials infrastructure is currently in place, and will probably remain so as long as it is competitive.

On the other hand, if all of panel manufacturing moves to Asia, it is unlikely that these materials companies can remain strong. The materials supply chain needs a base of panel manufacturing in the US to maintain its strength.

As for the equipment sector, much of the semiconductor equipment industry is located in Silicon Valley, including Applied Materials-TEL, Lam-Novellus, and KLA-Tencor. The risk is the same, though, in that the equipment supply chain often moves to where its panel-manufacturing customers locate. This is already happening. Most solar panel manufacturing has moved to China, and equipment companies in China are copying European and American designs

(Already, about 80% of laminators for solar panels are made in China). But Solar 2.0 may require a whole new set of tools. Innovation is the key. The equipment sector may, in fact, be wide open, and the US has the innovative power to maintain strength.

A Manufacturing Strategy for Solar 2.0

Can solar panel manufacturing be done competitively in the United States? Yes. The US has an abundance of technical and business advantages that apply to the manufacture of solar panels. We have a good physical infrastructure, low-cost and reliable electricity, and a dependable communications network. It also has good access to capital, a stable monetary system, dependable business practices, a quality supply chain, and a strong technology base. Human capital is another strength. The US is home to some of the best research scientists, engineers, and management in the world—and this talent goes deep. According to the US Bureau of Labor Statistics, in 2011 (the last year reported), the US ranked third in the world on productivity per capita, behind only Norway and Singapore.[71]

A subtler infrastructure issue is the rule of law. In the US, one can trust contracts and build a business around laws that are known and enforced; intellectual property is protected. Some low-cost areas of the world are weak in this regard. So, while countries with low-cost labor and cheap land may have a manufacturing cost advantage, that advantage diminishes when considering these other factors.

One of the most profound advantages of the US is that it is the world's innovation capital. In *The Innovator's Dilemma*,[72] Clayton Christensen discusses this topic beautifully. It's the American Dream that anyone with the will and capability can make the next big thing happen. This cultural mindset makes for an independent-minded, creative workforce. In making the transition from Solar 1.0 to 2.0, this is a critical plus.

Success in panel manufacturing, though, will take more than leaning on this long list of advantages. With China's strength in c-Si manufacturing, it will take executing on the right strategy, including four major points:

Select the right technology...
at the right scale...
with the right business model...
using advanced manufacturing principles.

1. Select the right technology

The answer for the US is thin film. The US cannot compete with China head-on in silicon panel manufacturing. China has too much of a head start, and silicon isn't going to get to the cost goal of $0.40/watt anyway. Thin film, though, could be the future of Solar 2.0. The US is home to First Solar, the best thin film company in the world, and US VC firms have invested billions to create the next great thin film company. The key is to select the right thin film technology and take that to scale, as China did with c-Si. The US, through public and private investment, should support the three most promising thin film candidates (outlined in Chapter 11):

1. CdTe on a larger substrate (1.6 – 2.0m^2) to be processed at the same yield and performance level as First Solar's smaller substrate.

2. Monolithic CIGS on glass: The market has already validated this technology, since most of the surviving CIGS companies, including Solar Frontier, Hanergy, TSMC, Stion, and Siva Power, all use monolithic CIGS on glass. So far, though, no one has been cost-competitive with either First Solar or silicon panels. Therefore, just like option #1, this technology needs to be executed on larger substrates, hopefully achieving the 300MW capacity outlined in the manufacturing roadmap in Chapter 13. If this can be accomplished, MLI CIGS on glass will be a big winner.

3. R2R CIGS: This will be a CIGS company that takes the best of what MiaSolé has done, and improves on it. That means eliminating diodes, going to a best-of-breed equipment architecture, and speeding up the tools — a lot.

If I had to pick one of these options, it would be monolithic CIGS on glass. In fact, to be honest, I had to pick one as CEO of Siva Power. We spent $60 million investigating CdTe, GaAs, CZTS, InP, and CIGS. We also investigated R2R vs. MLI, even building a pilot line for R2R CdTe on a flexible stainless-steel substrate. After years of research and development, we picked MLI CIGS because the data, on both cost and efficiency, was conclusive.

Let's look at the efficiency data that drove that decision. In late 2013, thin film technology, specifically CIGS, surpassed mc-Si in laboratory efficiency for the first time, 20.8% vs. 20.4% (Fig. 19.1). It is a small difference but a significant one, because throughout history c-Si has been the efficiency leader, and mc-Si is the mainstay of silicon solar. High efficiency and the ability to scale have been mc-Si's key advantages.

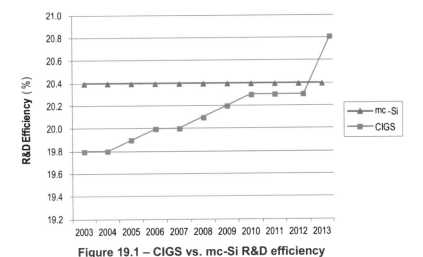

Figure 19.1 – CIGS vs. mc-Si R&D efficiency

This efficiency record, though, came in the lab; what really matters is what happens in production. It has not crossed over yet, but is gaining fast. The forecast, shown in Figure 19.2, is that CIGS will catch and surpass mc-Si in production within the next few years.

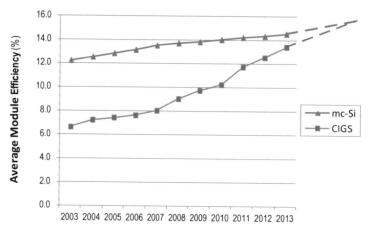

Figure 19.2 – CIGS vs. mc-Si HVM production efficiency

If this production efficiency crossover occurs, the prospect of thin film technology combining very high efficiency with very low manufacturing cost could come true. That low cost will come through "appropriate" scaling.

2. Scale like crazy, but not by replicating.
This is critical. Although thin film was once the leader in Solar 1.0, when China scaled c-Si by adding more than 40GW of capacity, thin film dropped to only 5% market share. Even though thin film is inherently less expensive, c-Si now beats most thin film companies on cost. With a current capacity of only 3GW, thin film has not scaled enough to match c-Si's impressive progress marching down the cost learning curve.

But…how to scale is important. Figure 19.3 shows how several successful industries have scaled.

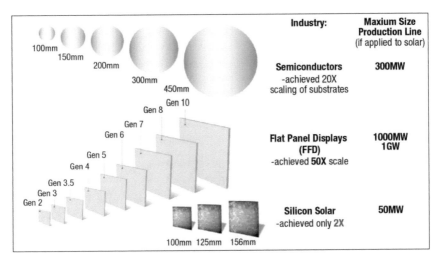

Figure 19.3 – Industry scaling of substrates

An important factor in scaling is the substrate size. The substrate is the material that these semiconductor devices are built on. In semiconductors and flat panel displays (FPDs), the substrates were scaled by 20X and 50X, respectively. What c-Si solar has done with only 2X pales in comparison. Silicon solar hasn't scaled more because it can't. The multi-crystalline wafers used as substrate, are delicate in nature and break when made much larger. Small substrates have limited the size of c-Si production lines to around 25MW. Silicon solar has scaled by just replicating these small lines -- over a thousand of them. But that can only work so far. At a certain point, further replication does nothing to reduce cost.

Other industries have scaled by building higher-volume production lines. If silicon solar could scale the way semiconductors or FPD scaled, the production lines could be 300MW to 1GW…on a single line. This is meaningful because of the known equipment scaling advantage; with larger substrates, every doubling of factory output only increases the capex by 20% to 30%. These economies of scale are shown in Figure 19.4 for hypothetical production lines going from 25MW to 1000MW in scale.

As shown, scaling to multi-hundred MW production lines has a much bigger impact on production costs than increasing efficiency.

Of course, one cannot arbitrarily build any size production line; if the high-throughput tools are not available, you're out of luck. Production line scaling therefore is limited by tool availability.

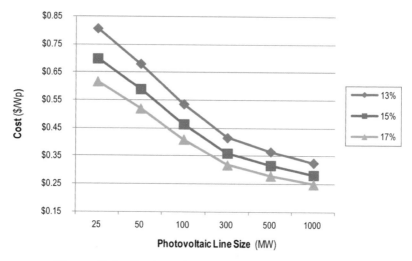

Figure 19.4 – Cost vs. "proper" scaling and efficiency

So the benefits of scaling through large substrate, high-capacity production lines is very clear, and this scaling is most readily accomplished not on silicon wafers but on large glass substrates. Fortunately, this is the substrate of choice for most thin film technologies.

Basically, silicon has just about scaled to its limit, but thin film is only getting started. But what about the "tool availability" question? Fortunately for thin film deposited on glass, an entire equipment set from the FPD and architectural glass coating industries has already been scaled and can be readily adapted to the solar industry. Thin film can build a production line of hundreds of MW as opposed to c-Si's 25-50MW limitation.

3. Select the right business model
Once the right technology is selected and scaled, the work has just begun. Learning from the billions spent so far, here are some best practices that all thin film companies should adopt.

Understand the business model. Thin film is different from silicon. Following the business model outlined in Chapter 15, skills and management talent are needed in both the materials and equipment areas. The need for experience with high throughput equipment to enable the scaling strategy is especially important. The successful company will also combine semiconductor and solar experts to improve execution.

Get good management. Thin film solar, because of the need for device, equipment, and materials expertise, is a tough business. The best management is required. With the industry going through dramatic consolidation, there is a lot of talent available to assemble a solar dream team. Do it.

Set the bar high. Several thin film companies failed even though they faithfully executed their plan. What they failed to recognize was that they could not win by merely copying the leader. They had to advance the state-of-the-art, to be 50% or even 100% better than the leader.

Be "glocal." The solar industry will be both global *and* local. The leaders in solar will export panels around the world, but they will also manufacture panels around the world, due to both economic realities and government policies.

4. Focus on advanced manufacturing, not new science
The key decider of success will not be which material is chosen, but which manufacturing process. Here is a short list of advanced manufacturing principles which should be implemented:

Automation is the answer: Automation can speed up processing, reduce labor costs, and improve quality. The labor that is required will be highly skilled engineers and technicians rather than machine operators. This is a perfect fit for the United States, as we have both the innovative equipment companies to design and build the automated tools, and the skilled labor force to support them. By reducing the advantage of low-cost labor, automation is a big step toward cost-competitiveness. This will reduce job creation, but there will be no jobs at all if the industry is not competitive.

Innovate in operations: The creativity usually reserved for R&D needs to be fostered in operations. The US has developed some of the most advanced operation methods in the world. Manufacturing innovations, like just-in-time inventory, unit metrics, and statistical process control, which have succeeded in other industries need to be applied to solar.

Chose process simplicity over efficiency: The attainment of world records in efficiency at the cost of process complexity makes no sense. Solyndra, Nanosolar, and other high-profile companies failed because of process complexity. Keep it simple; reduce the number of manufacturing process steps.

Invest in new equipment: The Solar 2.0 thin film factory of the future cannot be built with the tools that exist today. The most successful thin film companies either make or design their own equipment, but a whole new batch of highly automated, high-throughput systems is needed.

Close the gap: The gap in efficiency between the lab and the production floor is larger in thin film than in silicon technology. Closing this gap will bring thin film efficiencies on par with silicon and will lead to low-cost leadership.

Bring it All Together
With the right technology at the right scale, thin film will have both higher performance and lower manufacturing cost. By executing on the right business model using advanced manufacturing techniques, this thin film "factory of the future" can operate competitively in the United States.

Thin film, then, is the future, but c-Si has a big head start. Can thin film catch up and win out? Let's explore this question using cost as the metric for success.

Adopting the SunShot goal, we are looking for a selling price of $0.50/watt to achieve global grid parity. Allowing enough profit for a healthy, high-growth company, this translates into a manufacturing cost of $0.40/watt. As the forecast presented in Chapter 13 suggested and NREL data supports (Fig. 19.5), c-Si might get to a $0.50 cost, but is unlikely to get to the $0.40/watt level. On the other hand, at just 15% efficiency, a 300MW thin film production line could easily achieve a cost of less than $0.40/watt (Fig. 19.4). This allows for a $0.50/watt sales price with a profit margin that ensures sustainability.

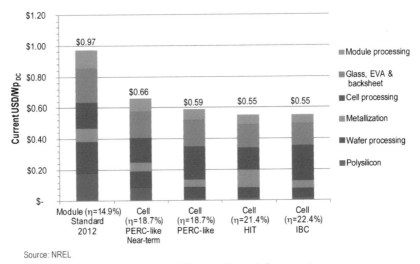

Source: NREL

Figure 19.5 – NREL mc-Si cost forecast

When China entered the solar market, it went into silicon not because it was inherently superior, but because it could easily be scaled through massive replication of existing technology. This strategy worked because there was a mature set of materials and equipment infrastructure available. All a manufacturer needed to do was to buy a factory and start churning out panels. That is how China built so much capacity, so quickly. But China is weak in thin film. In fact, no one has ever built a thin film factory competitive with First Solar. They control their equipment; because competitors cannot buy the equipment, they cannot build a factory.

There is a huge opportunity for American companies to follow First Solar's successful business model and build the thin film factory of the future. Because it is very hard to develop equipment, most solar manufacturers just buy their factory tools. This is why equipment expertise is so important for thin film companies. Sure, that makes the business model more difficult. But First Solar has proven the merits of this approach by being the most profitable manufacturer in solar.

A Matter of Will

The US has the technology, the talent, the innovative spirit, and the infrastructure to be successful in panel manufacturing. But the big question is: Do we have the will? The US has been challenged by overseas competitors before, and has risen to those challenges. Just look at computers, automobiles, airplanes, and consumer electronics to name a few.

History shows that when the United States decides to compete, it does. The semiconductor industry offers an interesting example. In the 1980s Japan was on the rise, and many feared they were going to take over the semiconductor industry. Then the industry and the government took action. Companies increased R&D and built advanced facilities all around the world. The US government helped to create Sematech, which brought together industry and public-sector leaders to strengthen the country's semiconductor industry. Together, they created a roadmap to guide the billions of dollars in R&D into a common direction. It worked. Today, the US has a tough and thriving semiconductor industry. Sure, some companies didn't make it. That's business. But Intel, Micron, and TI are all leading semiconductor firms, and Applied Materials is the world leader in semiconductor equipment.

The challenge in solar, given China's dominant starting position, may be unlike anything we have seen before. China has an intelligent and industrious workforce, a supportive government, and abundant capital. It is formidable. What confuses me is why America hasn't woken up to this challenge. Are we content to just develop iPhone apps, and leave manufacturing to others?

After the Cold War, we apparently lost something very important...a unifying national cause. Lacking an external focus, but with our strong competitive spirit, we have descended into fighting amongst ourselves, as evidenced by our current partisan and dead-locked congress. This is not only a waste of time and talent, but it is distracting the country from its real challenge: economic competitiveness.

Solar is leaving its infancy, and that makes this a fascinating time. In any transition, there are opportunities for new players. In fact, the newbies often dominate the second wave because they see things the older generation does not (think of Google beating out AOL in search, and Yahoo, Apple, and Samsung displacing Motorola and Nokia in cellphones). That is the power of thinking differently, of innovation. If thin film emerges as the market-leading technology, the US can capitalize on its strength in innovation, and regain leadership in Solar 2.0.

This is not a matter of technology, or policy, or even of money… it is a matter of will.

So… there it is.
The stakes are immense.
Do we have the will?

I say **YES!**

In my mind there is no question. We have the will and the way. The American people are a sight to behold when we decide to do something. This is simple compared to landing someone on the moon.

Writing this book is my way of contributing to the effort; drawing a picture of what success looks like. Let's engage that great American asset: the drive, the desire to succeed. When it is engaged, no one, *no one* can stop us. We've seen it happen before. We will do it again.

We *will* succeed!

Resources for Solar Information

Books:

Creating Climate Wealth (Jigar Shah)

Rooftop Revolution: How Solar Power Can Save Our Economy-and Our Planet-from Dirty Energy (Danny Kennedy)

Solar Electricity Handbook (Michael Boxwell)

Solar Revolution: The Economic Transformation of the Global Energy Industry (Travis Bradford)

Industry News:

Greentech Media (www.greentechmedia.com)

Renewable Energy World (www.renewableenergyworld.com)

PV Magazine (www.pv-magazine.com)

PV Tech (www.pv-tech.org)

Organizations:

American Solar Energy Society (www.ases.org)

European Photovoltaic Industry Association (www.epia.org)

Fraunhofer ISE (www.ise.fraunhofer.de/en)

International Energy Agency (www.iea.org)

Intersolar (www.intersolar.de/en)

National Renewable Energy Laboratory (www.nrel.gov)

Office of Energy Efficiency and Renewable Energy (www.energy.gov/eere/renewables/solar)

Solar Energy Industries Association (www.seia.org)

Organizations: (continued)

EC Joint Research Centre, Institute for Energy and Transport
(www.re.jrc.ec.europa.eu/pvgis/)

PV Cycle (www.pvcycle.org.uk)

Other Websites:

European Photovoltaic Technology Platform
(www.eupvplatform.org)

Green Building Advisor (www.greenbuildingadvisor.com)

Acronyms & Abbreviations

a-Si	amorphous silicon
AC	alternating current
ASP	average selling price
BOS	balance of system
BP	British Petroleum
BTU	British thermal unit
C&F	concept and feasibility
CdTe	Cadmium Telluride
CapEx	capital expenditure
CO_2	carbon dioxide
COGS	cost of goods sold
COO	cost of ownership
c-Si	crystalline silicon
CIGS	Copper Indium Gallium diSelenide
CZTS	Copper Zinc Tin diSelenide
DC	direct current
DG	distributed generation
DOE	Department of Energy
DRAM	dynamic random access memory
DNI	direct normal irradiation
DSIRE	Database of State Incentives for Renewables and Energy
EIA	Energy Information Administration
EPC	engineering-procurement-construction
FBR	fluidized bed reactor
FERC	Federal Energy Regulatory Commission
FIT	feed-in-tariff
FPD	flat panel display
GaAs	Gallium Arsenide
GDP	gross domestic product
GTM	Greentech Media
GW	gigawatt
HVM	high volume manufacturing
IEA	International Energy Agency
IEEE	Institute of Electrical and Electronics Engineers
IP	intellectual property

IPO	initial public offering
LBNL	Lawrence Berkeley National Laboratory (also LBL)
mc-Si	multi-crystalline silicon
MLI	monolithic integration
MLP	master limited partnerships
MWT	metal wrap-through
MOCVD	Metalorganic Chemical Vapor Deposition
NEC	National Electric Code
NP	nanoparticle
NREL	National Renewable Energy Laboratory
OPEC	Organization of Petroleum Exporting Countries
PACE	Property Assessed Clean Energy
PECVD	Plasma Enhanced Chemical Vapor Deposition
PERC, PERL	passivated emitter & rear cell, passivated emitter & rear locally diffused cell
Poly	polysilicon
PR	public relations
PUC	Public Utilities Commission
REIT	Real Estate Investment Trusts
RMI	Rocky Mountain Institute
ROI	return on investment
R2R	roll-to-roll
RTP	Rapid Thermal Processing
SC	singulated cells
SCI	Solar Cells Incorporated
SEIA	Solar Energy Industries Association
SMART	Specific Measurable Achievable Resourced Timed
SOC	silicon-on-cheap
SPE	Special Purpose Equity
SREC	Solar Renewable Energy Credits
SWOT	strengths-weaknesses-opportunities-threats
T&D	transmission and distribution
TPO	third party operator
UL	Underwriters Laboratory
VC	venture capitalist
VOST	value-of-solar tariff

Glossary of Terms

AC (alternating current) - Flow of electricity that reverses direction at a certain frequency. Common household and commercial electricity.

a-Si (amorphous silicon) - A common thin film solar material made of silicon with no crystal structure. It absorbs sunlight much better than crystalline silicon.

Antireflective coating (ARC) - A coating on solar cells or panels that prevents reflection, increasing efficiency by allowing more light to be absorbed by the solar cell.

Array - Combination of several individual solar modules connected to increase power output.

Back contact - Electrical contact at the back of a solar cell. Examples include silver or aluminum for silicon solar cells and molybdenum for CdTe and CIGS solar cells.

Band gap - Energy difference between the valance band and conduction band in a semiconductor material.

Breakdown voltage - Voltage in a semiconductor diode in the reverse direction that causes damage to the solar cell.

Carbon credit - See Renewable Energy Credit (REC)

CdTe (Cadmium Telluride) - CdTe is a thin film solar cell material. Often referred to as a II-VI (2-6) based on the position of Cd and Te in the periodic table. CdTe is a direct band gap semiconductor with a near ideal band gap (of 1.47 eV) to absorb sunlight.

CIGS (Copper Indium Gallium diSelenide) - CIGS is a thin film solar cell material. It is a direct bandgap semiconductor with a variable bandgap depending on the ratios of C-I-G and S. CIGS has the highest thin film solar cell efficiency, on par with mc-Si.

Conduction band - At the atomic level, the energy band where an electron is free to move, not bound to an atom. In solar cell operation, when light is absorbed, an electron moves from the valance band (of an atom) to the conduction band and can become electrical current.

CPV (Concentrated Photo Voltaic) - CPV is similar to solar PV in that sunlight is converted directly into electricity. However, this technology adds lenses or mirrors to concentrate the sun's energy into smaller spots, reducing the amount of photovoltaic material required.

CSP (Concentrated Solar Power) - CSP and CST (Concentrated Solar Thermal) have the same meaning. CSP technology concentrates the sun's energy with mirrors to heat fluids and generate steam. Much like geothermal, the steam drives a turbine in order to generate electricity.

DC (direct current) – Electrical current that flows in one direction, as typically found in batteries and solar cells.

DG (distributed generation) - Localized generation of electricity where the generation and use locations are in close proximity. DG sources are typically much smaller in capacity (expressed in watts, kW, or MW) than centralized electricity generation.

Dielectric material - Material that does not conduct electricity because electrons are pinned to specific atoms and cannot move freely.

Diode - An electrical device, in most cases a semiconductor p-n junction, that conducts electrical current in only one direction. A solar cell is a diode.

Direct bandgap - A PV material that absorbs sunlight very efficiently, needing only 1-2 microns to absorb almost 100% of solar energy.

Efficiency - Efficiency (or conversion efficiency) is the ratio expressed as a percentage of electrical energy to incident energy of incoming sunlight (international convention of 1000 W/m^2).

Electrons - A subatomic particle, symbol e-, with a negative charge. Electrons are the primary charge carrier in metals and semiconductors, and their movement creates electricity.

Encapsulation - Used in a solar panels, the encapsulant fills in and "glues" the solar cells between a front sheet (usually glass) and a back sheet (glass or a polymer moisture barrier). Common encapsulants are ethylene vinyl acetate (EVA), polyvinyl butyral (PVB), and olefins.

FIT (feed-in-tariff) - A FIT is a government policy that incentivizes investment in renewable energy by assuring advantageous long-term payments (tariff) for renewable energy generation.

Fill Factor (FF) - The ratio of the product of the maximum power point voltage and maximum power point current over the product of the maximum open circuit voltage and short circuit current: $(V_{mp} \times I_{mp}) \div (V_{oc} \times I_{sc})$.

Fossil fuels - Naturally occurring carbon (coal) or hydrocarbon (petroleum, natural gas) formed by the decomposition of prehistoric plants and animals.

Fracking - Fracking (hydraulic fracturing) is a process of injecting high pressure fluids into rock formations, causing cracks to grow and spread, releasing more oil or natural gas than would naturally be released.

Frequency – With electricity, frequency is the number of cycles of AC current.

Front contact – The front electrical contact on a solar cell, e.g. silver grids for silicon solar cells and transparent conducting oxides for thin film solar cells.

Global warming - See Greenhouse effect.

Greenhouse effect - Gaseous molecules in the atmosphere absorb sunlight and raise the temperature of the atmosphere. Greenhouse gases include methane, CFCs, and CO_2. The largest source is CO_2, created by the burning of fossil fuels.

Grid-connected - Power sources that are connected to the electrical grid.

Grid parity - The point at which an energy source can produce power at the same cost as grid electricity.

Indirect Bandgap - A PV material, like crystalline silicon, where the valance band maximum does not align with the conduction band minimum. The misalignment requires more energy for an electron to jump from the valance band to the conduction band. As a result, this material does not absorb sunlight very efficiently, taking 100-200 micron (100X more than direct bandgap) to absorb sunlight.

Ingot –A cast block of multi-crystalline silicon or a pulled boule of single-crystal silicon that is later sawed into wafers.

Insolation - A measure of the solar energy received by a given surface during a given time. Insolation maps allow designers of solar power plants to predict the power output over time.

Insulator – See Dielectric materials.

Interconnection - Process of electrically interconnecting solar cells to form a solar panel. In silicon cells, interconnection is typically accomplished with ribbons. In thin film, interconnection is typically accomplished by monolithic integration (MLI).

Inverter - A device that converts DC electricity into AC electricity.

Investment Tax Credit (ITC) – Provides an incentive to install renewable energy by providing a tax credit for the installer or owner.

I-V Curve - Plot showing the relationship between current and voltage in a solar panel or cell. The I-V curve provides key performance characteristics including open circuit voltage (V_{oc}), short circuit current (I_{sc}), fill factor (FF), maximum power point voltage (V_{mpp}), maximum power point current (I_{mpp}) series resistance (R_s), and shunt resistance (R_{SH}).

Kerf (Kerfless) - Kerf is the silicon lost (as sawdust) when sawing a boule or ingot into wafers. Kerfless technologies are when wafers are fabricated without sawing.

kW (kilowatt) = 1000 watts

kWh (kilowatt-hour) – kWh is the measure for electricity, when a kilowatt is used or created for one hour. An average U.S. home uses about 32 kWh per day.

Levelized Cost of Energy (LCOE) –A formula used to calculate the total cost to generate electricity, allowing comparisons of different energy sources (e.g., coal, solar, wind, hydro, nuclear).

Maximum power point (MPP) - Abbreviated either mp or mpp, maximum power point is the point on the solar panel or solar cell I-V curve where the product of current and voltage is maximized.

mc-Si (multi-crystalline silicon) – mc-Si is composed of multiple large grains (~1cm) of single crystal silicon. It has crystal structure but not as perfect at single-crystal silicon and therefore lower in efficiency. The mc-Si term refers to the material itself and also to solar cells or modules made out of the material.

Megawatt (MW) - MW = 1,000,000 watts or 1,000 kWs.

Megawatt-hour (MWh) – MWh is equal to a MW of energy used or created for one hour. An average home consumes about 12 MWhs of electricity per year.

Micron (μm) - Is a thickness measurement = micrometer = 1/1,000,000 of a meter.

Module - See Solar Module.

Monolithic Integration (MLI) - An approach to create a series interconnection between adjacent thin film solar cells using automated laser or mechanical scribing. The scribes are used to isolate the front and back contacts of adjacent.

Multi-Junction Cell - Multi junction solar cells are cells with multiple diodes or p-n junctions. Each p-n junction is designed to absorb a specific portion of incoming sunlight. The top cells (large band gap) and absorb high energy short wavelength light (Purple, blue, green) and the bottom cells (small band gap) absorb lower energy long wavelength light (red, near infrared).

n-type - A semiconductor that is doped or modified in a way to have an excess of electrons.

Negawatts - Negawatts in the amount of energy (in watts) that is saved or not used by implemented energy conversation or improving energy efficiency.

Net Metering - Net metering is an incentive program implemented by utility companies that allows electricity generated at a site, on the customer side of the meter, to offset the electricity supplied to that site at other times, essentially allowing the meter to run backwards.

Open Circuit Voltage (V_{oc}) - The voltage seen when there is no load, therefore no current flow from a source of voltage. In solar, it is the voltage on the I-V curve where I_{sc} is zero.

p-n Junction - p-n junction is the metallurgical interface or boundary between a p-type and an n-type semiconductor. p-n junctions create a diode allowing current flow in only one direction, and also create both solar cells and LEDs, light emitting diodes.

p-type - A semiconductor that is doped or modified in a way to have a deficiency of electrons or stated another way, an excess of holes.

Passivation - A process used in solar cells to tie up or "passivate" what would otherwise be open bonds that trap electrons and reduce efficiency.

Photon - An elementary particle (quantum of light) that has no mass or electrical charge. Photons are absorbed by solar cells to create electricity.

Photovoltaic (PV) - Semiconductor devices or materials that convert photons (light) into electrical energy.

Photovoltaic Effect - When photons are absorbed by PV materials and electrons move from the valence band to the conduction band, creating the possibility for voltage and electrical current (electricity).

Production Tax Credit - A production tax credit is a per-kilowatt-hour credit for electricity generated by qualified energy resources and sold by the taxpayer to an unrelated person during the taxable year.

Purchase Power Agreement (PPA) - An agreement between an electrical energy provider (seller) and electrical energy purchaser (buyer) that defines the terms of sale between the two parties, usually over a long period of time, providing a predictable cash flow and therefore incentive to invest.

Radiation - A process in which energetic particles or energy waves travel though a media. Types of radiation include sound, heat, microwaves, UV, and light.

Recombination - In solar, recombination is a process where electrons that were transferred from the valance band to the conduction band fall back into a hole in the valance band and as a result do not participate in the creation of electricity.

Renewable Energy Credit (REC) - Environmental commodity certificates that represent the added value of a renewable energy system. Organizations that produce renewable energy can sell the REC's to organizations that have a REC appetite including utilities companies and businesses or agencies that wish to be environmentally responsible but do not have the means of producing renewable energy.

Renewable Portfolio Standards (RPS) - A government imposed regulation that requires a certain portion generated power to be from renewable sources such as solar, wind, biomass, or geothermal.

Reverse Breakdown - See Breakdown Voltage

sc-Si (single-crystal silicon) - sc-Si or monocrystalline silicon is a large single crystal of silicon. It has fewer defects (that contribute to recombination) and therefore the highest efficiency of any silicon material, but is also the most expensive.

Semiconductor - A material that exhibits electrical conductivity somewhere between a conductor (a metal) and an insulator (a ceramic, glass, or plastic). Common semiconductors include silicon, III-Vs like GaAs, II-VIs like CdTe and CIGS. They are used to form diodes (LEDs and solar cells), transistors, and ICs.

Series Resistance (R_s) - Resistance to the flow of electricity through the solar cell structure. The main sources of series resistance are at the interfaces: resistance at the p-n junction and at the front (top) and back contacts. R_s reduces efficiency therefore for an ideal solar cell R_S should be zero.

Short Circuit Current (I_{sc}) - Short circuit current is the free flowing current when a circuit is completely shorted. For solar cells, I_{sc} is the current through the solar cell with the voltage across the solar cell is zero ($V_{oc} = 0$).

Shunt Resistance - Shunt resistance (R_{SH}) is the resistance to current flow though a solar cell through pathways other than the intended path. Shunting is usually caused by macro-level defects in the solar cell device introduced during manufacturing. For an ideal solar cell, R_{SH} should be infinity.

Silicon (Si) - Silicon is the world's most important semiconductor material.

Solar cell (or photovoltaic cell) - An electrical device that produces DC electricity from sunlight.

Solar module or panel - A set of solar cells that are electrically interconnected and laminated with encapsulant to form a module.

Solar spectrum - The solar spectrum is a quantification of the solar irradiance intensity as a function of wavelength of the solar irradiation.

Standard test condition – Established by the international solar community to standardize measurement. (E.g. efficiency is measured at $1000W/m^2$ incident energy at AM1.5 and 25°C solar cell surface temperature.)

Substrate – A base upon which processes are done. In solar it refers to the base the solar cell is built on. In c-Si technology, it is a silicon wafer. In thin films, it could be glass, stainless steel, aluminum, or plastic.

Superstrate - Superstrate is similar to a substrate with the exception that with superstrate the solar cell is built upside down. With substrates, the last layer deposited is the layer exposed to the sun. With superstrate the first layer is the one exposed to the sun, therefore the superstrate material must be transparent to allow the light through to the solar cell.

Tandem (Tandem Junction) - A tandem junction cell is a multi-junction cell with only two p-n junctions.

TCO (Transparent Conductive Oxide) - A material that is both transparent and electrically conductive which is used for the top contact on thin film solar cells.

Thin film - A PV material that is direct bandgap and can therefore can be very thin and still absorb most of the sunlight. Thicknesses range from 1-6 microns. Common thin film PV materials are: a-Si, CdTe, CIGS, and GaAs.

Utility-scale - Utility-scale solar is a large (~10-1000MW) centralized power facility that delivers electricity to the grid for transmission and distribution.

Wafer - The substrate for c-Si solar cells, but also used to fabricate semiconductor ICs.

List of Illustrations

Chapter 17

Figure 17.3 – Electricity use per person US vs. California

Figure 17.4 - Historical average of annual federal energy subsidies

Figure 17.5 – Current US status of incentives and subsidies

Chapter 19

Figure 19.1 – CIGS vs. mc-Si R&D efficiency

Figure 19.2 – CIGS vs. mc-Si HVM production efficiency

Figure 19.3 – Industry scaling of substrates

Figure 19.4 – Cost vs. "proper" scaling and efficiency

Figure 19.5 – NREL mc-Si cost forecast

Additional Illustration Source Information

Figure 1.2 - Atmospheric CO_2 vs. surface temperature
Source: Brohan, P., Kennedy, J.J., Harris, I., Tett, S.F.B., & Jones, P.D. (2006) Uncertainty estimates in regional and global observed temperature changes: A new data set from 1850. *Journal of Geophysical Research: Atmospheres (1984-2012), 97* (D12).

Figure 8.4 – Learning curve, high volumes drive lower costs
Sources: Paul Maycock, Bloomberg New Energy Finance, and FSLR filings.

Figure 8.5 - Top 10 global solar-panel manufacturers (by volume)
Sources: European Commission (2014), IC Insights (2008), Solarbuzz (2012).

References

Chapter 1

[1] Cornell University (2007) Pollution Causes 40 Percent of Deaths World-wide, Study Finds. Science Daily. Retrieved from www.sciencedaily.com/releases/2007/08/070813162438.htm

[2] WHO (2011). GHO: *Deaths attributable to outdoor air pollution*. World Health Organization, Department of Public Health Information and Geographic Information Systems.

[3] Sachs, J. and Warner, A.M. (1997) *Natural Resource Abundance and Economic Growth*. Center for International Development and Harvard Institution for International Development. Harvard University. Cambridge. MA.

[4] GFDL (1999). *Climate Impact of Quadrupling CO2*. Retrieved from URL http://gfdl.noaa.gov/index/knutson-climate-impact-of-quadrupling-co2

[5] Krijgsman, R. (2014). *Shell's Profit Warning a Symptom of Wider Trend Among the Majors*. Retrieved from http://blog.evaluateenergy.com/shells-profit-warning-a-symptom-of-wider-trend-among-the-majors

[6] BP Worldwide (2013). *Statistical Review of World Energy 2013*. Retrieved from URL http://www.bp.com/en/global/corporate/about-bp/energy-economics/statistical-review-of-world-energy-2013

Chapter 2

[7] Wikipedia (2014). *List of countries by electrical consumption*, Retrieved from http://en.wikipedia.org/wiki/List_of_countries_by_electricity_consumption

[8] Independent analysis based on energy gap of 40 billions of barrels in 2050, 1628.2 kW-hr/barrel, 77% STC_{DC} Installed vs. Grid AC power conversion, and average solar day of 5 hours. *40 billions of barrels*. Retrieved from SRI Consulting. *Billion Barrel to kW-hr*. Retrieved from http://www.unitjuggler.com/convert-energy-from-kWh-to-Mboe.html. *77% STC_{DC} Installed vs. Grid AC power conversion*. Retrieved from NREL PVWatts

http://rredc.nrel.gov/solar/calculators/pvwatts/system.html. and *Average Solar Day* Retrieved from http://www.hotspotenergy.com/DC-air-conditioner/usa-solar-hours-map.php

[9] Office of Energy Projects (2013). *Energy Infrastructure Update.* Federal Energy Regulatory Commission (FERC). Retrieved from https://www.ferc.gov/legal/staff-reports/2013/jan-energy-infrastructure.pdf.

[10] Nuclear Energy Institute (2014). *Nuclear Units Under Construction.* Retrieved from http://www.nei.org/Knowledge-Center/Nuclear-Statistics/World-Statistics/Nuclear-Units-Under-Construction-Worldwide

[11] BP (2013). *BP Statistical Review of World Energy.* Retrieved from http://www.bp.com/content/dam/bp/pdf/statistical-review/statistical_review_of_world_energy_2013.pdf. Page 41.

Chapter 3

[12] EIA FAQ (2012). *How Much Electricity does and American Home Use?.* U.S. Energy Information Administration. Retrieved from http://www.eia.gov/tools/faqs/faq.cfm?id=97&t=3

[13] Climate Investment Funds (2014). *SREP Investment Plan for Solomon Islands.* Retrieved from https://www.climateinvestmentfunds.org/cif/sites/climateinvestmentfunds.org/files/SREP_11_5_Solomon_Islands_IP_final.pdf

[14] EIA FAQ (2012). *How Much Electricity does and American Home Use?.* U.S. Energy Information Administration. Retrieved from http://www.eia.gov/tools/faqs/faq.cfm?id=97&t=3

[15] Perez, R., Kmiecki, M., Hoff, T., Williams, J.G., Herig, C, Letendre, S., and Margolis, R.M. (2004). Availability of dispersed photovoltaic resource during the August 14th 2003 Northeast Power Outage. *U.S. Department of Energy (U.S.) DOE) Office of Energy Efficiency and Renewable Energy (EERE) under National Renewable Energy Laboratory (NREL) Contract No. AAD-2-31904-01.*

[16] Baughman, M.L., Bottaro, D.J. (1976). Electrical Power Transmission and Distribution Systems: Costs and Their Allocation. *Power Apparatus and Systems, IEEE Transactions. 95(03)*, 1.

Chapter 4

[17] Lewis, N.G. and Nocera, D.G., Powering the planet: Chemical Challenges in Solar Energy utilization", *Proceedings of the National Academy of the Sciences, 103 (43),* 15729–15735.

[18] Giles Parkinson (2013). *Deutsche Bank: Solar, distributed energy at 'major inflection point',* RE New Economy. Retrieved from http://reneweconomy.com.au/2013/deutsche-bank-solar-distributed-energy-at-major-inflection-point-10487

Chapter 5

[19] EIA (2013). *Updated Capital Cost Estimates for Utility Scale Electricity Generating Plants.* U.S. Energy Information Administration. U.S. DOE, Washington, DC.

[20] Branker, K., Pathak, M.J.M., Pearce, J.M. (2011). *A review of solar photovoltaic levelized cost of electricity,* Renewable and Sustainable Energy Reviews, *15(2011),* 4470-4482.

Chapter 6

[21] Shiao, M.J., (2012). *Thin Film Manufacturing Prospects in the Sub-Dollar-Per-Watt Market, What Happened to the promise of thin-film PV?* Greentech Solar, Retrieved from . http://www.greentechmedia.com/articles/read/thin-film-manufacturing-prospects-in-the-sub-dollar-per-watt-market

[22] Watanabe, C., (2013). *Japan Banks Ante Up to $19 Billion Solar Market,* Renewable Energy World. Retrieved from http://www.renewableenergyworld.com/rea/news/article/2013/02/japan-banks-ante-up-to-19-billion-solar-market

Chapter 7

[23] Staff of the Joint Commission on Taxation (1977). Committee on the Ways and Means, House of Representatives, US Government Printing Office.

Chapter 8

[24] Mehta, S., (2013, August). Where Will PV Module Costs Bottom Out?. Greentech Media Webinar. Conducted from https://event.on24.com/eventRegistration/EventLobbyServlet?target=registration.jsp&eventid=668030&sessionid=1&key=6CC935C001CA11E63A5642E17638B677&partnerref=email2&sourcepage=register

[25] Wynn, G. (2014). *Why are solar panel prices starting to rise?*. Responding to Climate Change. Retrieved from http://www.rtcc.org/2014/06/02/why-are-solar-panel-prices-starting-to-rise/

[26] Ong, S., Denholm, P., Clark, N., (2012). *Grid Parity for Residential Photovoltaics in the United States: Key Drivers and Sensitivities*, World Renewable Energy Forum, Denver, CO, May 13-17, 2012

[27] Denholm, P., Margolis, R.M., Ong, S., Roberts, B. (2009). *Breakeven Cost for Residential Photovoltaics in the United States: Key Drivers and Sensitivities (Report Summary).* National Renewable Energy Laboratory, NREL/TP-6A2-46909.

[28] Breyer, C., Gerlach, A. (2012) Global Overview on Grid Parity. *Prog. Photovolt: Res. Appl., 21(01)* 121-136

Chapter 9

[29] Keiser, R., (2011) Retrieved from http://www.keiser-analytics.com/data.html

[30] Parkinson, G., (2013) *Deutsche Bank says US solar boom to reach 50GW by 2016*. RE New Economy, Data retrieved from http://reneweconomy.com.au/2013/deutsche-bank-says-us-solar-boom-to-reach-50gw-by-2016-18298

[31] Mehta, S. (2012). *Global PV Module Manufacturers 2013: Competitive Positioning, Consolidation and the China Factor.* GTM Research. Retrieved from http://www.greentechmedia.com/research/report/global-pv-module-manufacturers-2013

[32] Goodrich, A, Noufi, R., Whitney, W., Woodhouse, M. (2013, March). *Techno Economic Analysis: Overview of NREL Road Map activities, in Support of the PVMC*, Presented at PVMC, Albany, NY, National Renewable Energy Laboratory, Golden, CO.

Chapter 11

[33] Applied Materials (2011) . *Screen Printed Selective Emitter Formation and Metallization*. Applied Materials, Santa Clara, CA. Retrieved from http://www.appliedmaterials.com/sites/default/files/Screen-Printed-Selective-Emitter-Whitepaper_EN.pdf

[34] Keyes, B. (2007). *National Solar Technology Roadmap, Film-Silicon PV*. DOE, Office of Energy Efficiency and Renewable Energy (EERE). Management Report NREL/MP-520-41734.

[35] SEMI, ITRPV (2013). *International Technology Roadmap for Photovoltaic, Fourth edition*. SEMI PV Group. San Jose, CA.

[36] Widmar, M. (2013). *First Solar Financial Summary*. Retrieved from http://files.shareholder.com/downloads/FSLR/3067502677x0x735354/0f1a5708-08ca-430d-bc12-5f2b05af27b1/FS_AnalystDay_FinancialGuidance.pdf

Chapter 13

[37] Goodrich, A., Hacke, P., Wang, Q., Sopori, B., Margolis, R.M., James, T.L., Woodhouse, M. (2013). *A Wafer-Based Monocrystalline Silicon Photovoltaics Road Map: Utilizing Known Technology Improvement Opportunities for Further Reductions in Manufacturing Costs*. Solar Energy Materials and Solar Cells *114(2013)* 110-135.

[38] Same as 37

Chapter 14

[39] GTM Staff (2011). *PV BOS Cost Analysis: Ground-Mounted Systems*. Greentech Media. http://www.greentechmedia.com/articles/read/pv-bos-cost-analysis-ground-mounted-systems

[40] Data provided by Fred Tabrizi, Siva Power's Vice President and President of Sierra Solar Systems.

[41] Rocky Mountain Institute (2014). *Simple Solar Balance of Systems (BoS)*. Retrieved from http://www.rmi.org/simple

Chapter 15

[42] Day, R. (2012, July). *Lessons From the Past Ten Years: Overreaction. Cleantech Investing*, Retrieved from http://www.greentechmedia.com/cleantech-investing/post/lessons-from-the-past-ten-years-overreaction

[43] Denholm, P., Margolis, R. (2008). *Supply Curves for Rooftop Solar PV-Generated Electricity for the United States*. National Renewable Energy Laboratory, Technical Report NREL/TP-6A0-44073.

[44] SEIA (2014). Shared Renewables/Community Solar. Solar Energy Industries Association. Retrieved from http://www.seia.org/policy/distributed-solar/shared-renewablescommunity-solar

[45] Trabish, H.K., (2013). *Securitization: Another Innovation in Solar Finance*. Greentech Solar. Retrieved from http://www.greentechmedia.com/articles/read/Securitization-Another-Innovation-In-Solar-Finance

[46] Alafita, T., Pearce, J.M. (2014). *Securitization of residential solar photovoltaic assets: Costs, risks and uncertainty,* Energy Policy, 67(2014), 488–498.

Chapter 16

[47] Burijn, J. (2013). *Top 10 Solar PV Stocks over 2013*, Solar Plaza, Retrieved from http://www.solarplaza.com/article/top-10-solar-pv-stocks-over

Chapter 17

[48] EIA (2014). *U.S. Imports by Country of Origin*, U.S. Energy Information Administration, Retrieved from

http://www.eia.gov/dnav/pet/pet_move_impcus_a2_nus_ep00_im0_mbbl_a.htm

Chapter 17 (continued)

[49] EPA (2014). *Renewable Energy Certificates (RECs)*. United States Environmental Protection Agency, Retrieved by http://www.epa.gov/greenpower/gpmarket/rec.htm

[50] Solar Server (2014). *New Spanish solar policy to result in cuts in payment up to 45%,* Solar Server, Global Solar Industry Website, Retrieved from http://www.solarserver.com/solar-magazine/solar-news/current/2014/kw06/new-spanish-solar-pv-policy-to-result-in-cuts-in-payment-up-to-45.html

[51] Lynch, P.J., (2011). *Feed-in-Tariffs: The Road Proven Road not Taken...Why.* Renewable Energy World, Retrieved from http://www.renewableenergyworld.com/rea/news/article/2011/11/feed-in-tariffs-the-proven-road-not-takenwhy

[52] Coughlin, J., Grove, J., Irvine, L., Jacobs, J.E., Phillips, S.J., Sawyer, A., Wiedman, J. (2010). *A Guide to Community Solar, Utility, Private and non-profit Project Development,* NREL, US DOE/EERE.

[53] EDF (2014). *How cap and trade works, Environmental Defense Fund,* Retrieved from http://www.edf.org/climate/how-cap-and-trade-works

[54] Murphy, C., (2002) *Hog wild for pollution trading,* Fortune, *146(4)* 137-140.

[55] Carbon Tax Center (2014). Retrieved from http://www.carbontax.org/

[56] Powersaver™ Program (2013). *New Value of Solar Rate Takes Effect January (2014),* Retrieved from http://austinenergy.com/wps/portal/psp/about/press-releases/new-value-of-solar-rate-takes-effect-january/

[57] Farrell, J., (2014). *Could Minnesota's "Value of Solar" Make Everyone a Winner?,* Renewable Energy World, Retrieved from http://www.renewableenergyworld.com/rea/blog/post/2014/03/could-minnesotas-value-of-solar-make-everyone-a-winner

Chapter 17 (continued)

[58] PVTech, (2014). *Tariff Watch, PVTech.org.*, Retrieved from http://www.pv-tech.org/tariff_watch/list

[59] Zubrinich, P., (2013). *Global FIT Overview, PV Magazine, Photovoltaic Markets and Technology*, Retrieved from http://www.pv-magazine.com/archive/articles/beitrag/global-fit-overview-_100010396/#axzz2NYOLC8M4

[60] Ren21 (2014). *Renewables Interactive Map*, Renewables Energy Policy Network for the 21st Century, Retrieved from http://www.map.ren21.net/#rw

[61] Montgomery, J., (2013). *100 GW of Solar PV Now Installed in the World Today.*, Renewable Energy World, Retrieved from http://www.renewableenergyworld.com/rea/news/article/2013/02/100-gw-of-solar-pv-now-installed-in-the-world-today

[62] DSIRE[TM] (2014). *Database of State Incentives for Renewables and Efficiency*, U.S. Department of Energy, Energy Efficiency and Renewable Energy (EERE), Retrieved from http://www.dsireusa.org/

[63] Energy.Gov (2014). *Section 1705 Loan Program.* U.S. Department of Energy Loan Program Office. List retrieved from http://energy.gov/lpo/services/section-1705-loan-program

[64] Plumer, B. (2013). *Why do Californians use less electricity than everyone else?*, The Washington Post, Wonk Blog, Retrieved from http://www.washingtonpost.com/blogs/wonkblog/wp/2013/08/12/why-do-californians-use-less-electricity-than-everyone-else/

[65] Pfund, N., Healy, B., (2011). *What Would Jefferson Do? The Historical Role of Federal Subsidies in Shaping America's Energy Future*, DBL Investors, San Francisco, CA.

[66] Blackmon, D., (2013). *Oil & Gas Tax Provisions Are Not Subsidies For "Big Oil*, Forbes.com.

[67] The Vote Solar Initiative (2014). Retrieved from http://votesolar.org/campaigns/projectpermit/

[68] EIA (2012). Project *sponsors are seeking Federal approval to export domestic natural gas*, U.S. Energy Information Administration, Retrieved from http://www.eia.gov/todayinenergy/detail.cfm?id=5970

[69] EIA (2013). *AEO2014 Early Release Overview*, U.S. Energy Information Administration, Retrieved from http://www.eia.gov/forecasts/aeo/er/pdf/0383er(2014).pdf

Chapter 18

[70] Olson, S., (2011*) Survey by Kelton Research finds that 89% of Americans support use and development of solar energy*, PVTech.org, retrieved http://www.pv-tech.org/news/survey_by_kelton_research_finds_that_89_of_americans_support_use_and_develo

Chapter 19

[71] Bureau of labor Statistics Report (2012). *International Comparisons of GDP per Capita and per Hour, 1960-2011. BLS, Division of International Labor Comparisons*, Retrieved from http://www.bls.gov/fls/intl_gdp_capita_gdp_hour.pdf

[72] Christensen, C.M. (1997). *The Innovators's Dilemma: When New Technologies Cause Great Firms to Fail,* Harvard Business School Press, Cambridge, MA.

CPSIA information can be obtained at www.ICGtesting.com
Printed in the USA
LVIW01n2348200717
542106LV00018B/84